THEM

Unsimplifying Common Perceptions and Broadening Perspectives in "The Public Understanding of Science"

The standard metrics don't always tell the full story.
-John Matsui

Copyright © 2016 by mcbfifteen

All rights reserved. This book or any portion thereof may not be reproduced or used in any manner whatsoever without the express written permission of the publisher except for the use of brief quotations in a book review or scholarly journal.

First Printing: 2016

ISBN 978-1-365-06536-1

MCB15
Course at the University of California, Berkeley
Berkeley, CA 94720

http://mcbfifteen.tumblr.com/

ILLUSTRATIONS

Cover: Julie Mendoza
Chapter Introductions: Julie Mendoza
Chapter 2 and 5 Graphic Design: Sabrina Rentschler

EDITORS-IN-CHIEF

Danny Lee
Julie Mendoza
Vin Lay
Jessica Yescas
Sabrina Rentschler
Dax viviD
Michael Ma
Alexiz Gomez

Contents

INTRODUCTION 14

Chapter 1: Barriers 22

Part 1: How Barriers to Understanding are Barriers to Health 25

    Won Jae Jeong: What Does That Even Mean? 26
    Claudia Campos: In Need of Better Distribution of Knowledge 28
    Hien Do: How Barriers to Understanding are Barriers in Health 29
    Michael Ma: Cultural, Social, and Political Tides of Misunderstanding as a Threat to Health 33
    Vivian Liu: Language Barriers to Understanding Health 35
    Brian Woo: Personal Evaluation of Religion and Science 37
    Will Lopez: Education 38

Part 2: Underserved Communities 41

    Karen Rodriguez: The Lack of Access to Community Resources In Underserved Communities 42
    Tiffany Chang: Access to Nutrition and Education 44
    Reinere Jude Ruiz: Underrepresentation in Medicine 45
    Lidia Lisette Pureco: Social Workers 47
    Juan Cano: Como Geografia Es Un Factor Important Que Excluye Comunidades Subatendidas De Recursos Basicos 49
    Alicia Cuevas: Increasing STEM Education 52
    Thu-An Tran: Language Barriers and Health Care 55
    Anonymous: The Undocumented 59
    Dawa Gangshar: Improving Nutrition, Reducing Obesity 62

Part 3: Language and Cultural Barriers 66

    Glendy Ramon: Language and Cultural Barriers within Latino immigrants 67
    Joanna Ruelas: The Need for Diverse Health Care Providers in Latino Communities 68
    Maricarmen Hernandez: Language and Cultural Barriers: Mediating Barriers in Health Care Settings 70
    Anthony Meza: How Economic and Language Barriers Prevent Underserved Communities from Accessing and Engaging in Science 72
    Claudia Pacheco: Latino Faith and Medicine 74
    Ruvim Kushnir: Mistrust in the Community 77

Part 4: The Demand of Non-Scientists to "Understand Science" 81

    Jennifer Diaz: Medicine as a Key to Motivation 82
    Claudia Tse: Harmful Effects of Ignorance and Disinterest 84
    Emani Harris: Is Understanding Health Enough? 85
    Khalid Al-Rayess: The Role Social Media Plays 86
    Xin (Natalie) Yu: Social Pressure 87
    Mark Chang: The Incentives Behind Public's Demand of Science 89

Part 5: On Becoming a Scientist from a Non-Science Background .. 91

    Emily Diaz 92

    Leilani Gutierrez-Palominos 93
    Gila Juárez: The First Generation Perspective 95
    Tarun Mendoza 97
    Cristina Batarse: The Struggle 99
    David Lamb: A Narrative Reflection of Different Resources 102
    Siso Phouthavong: My Journey into the Sciences as a Pomo Indian 105

Chapter 2: Public Experiences and Perspectives in Science 111

Part 1: Changing Perspectives with Development 114
    Sophie Ballard 115
    Jacob Cota 116
    Betsy Rosales Avalos 118
    Keana Richmond 120
    Diane Chong 122
    Diana Bahena 125
    Vanessa Sarmiento 126

Part 2: Science, Family, and Community 130
    Mar Jean 131
    Wendi Ruiz 132
    Nancy Cuevas 135
    Rafael Mejia 136
    Adela Ramos 139
    Minh-Thy Nguyen 141
    Maria Rivera 142
    Marycon Jiro 145
    Nicholas Garcia 148
    Gabriela Wantah 149
    Rattana Sot 153
    Johannah Perez 154
    Paolo Joaquin: The Filipino Community and the Achy Breaky Heart 157
    Eliott Ahumada 159
    Andrew Wing 160

Chapter 3: Public Engagement of Science 162

Part 1: "What's in it for me?" 164
    David Barajas 165
    Jaewon (Evelyn) Lee: Science Education and Female Students of Color 166
    Miriam Juarez: Engaging Low-Income Communities in the Science of Climate Change 169
    Elizabeth Aguayo: Giving Communities a Voice 170
    Austin Weinstein: Science in American Political Speech 172
    Liang-Han Tai (Chase): Science as Performance: Problems and Solutions 174
    Steven Yu 177
    Josephine Espinoza 178

Part 2: "How can we engage?" 181

Nick Garelis 182
　　Deena Abdelhalim 183
　　Charles Li: It's A Two-Way Street: Beyond the Public Understanding of Science 185
　　Krystin Ventura: Young Women and Science During the Critical Period 188
　　Lorene Cudjoe: *A Focus on Middle School Girls*: Cultural Humility as the Key to Destigmatize Science 190
　　Jonathan Homidan: *Engage:* Media Intervention and Sociology 191
　　Sarah Cho: Interactive Methods, Mandatory Education 193

Chapter 4: Scientists are people, too: Critical Review of Science 198

Part 1: Ethical Concerns in Science 202
　　Diego DuBon: Ethical Concerns in Science: Medical Ethics 203
　　Joseline Padilla Alvarez: Ethical Aspects of Designer Babies 206
　　Joseph Jweinat: Is it Safe and Moral to Edit the Human Genome? 207
　　Jazmine Carvajal: Artificial Insemination: Is it Acceptable? 210
　　Anonymous: Prison Experimentation 211
　　Terry Kyubin Kyung: Moral Dilemma around Embryonic Stem Cell Research 213
　　Denisse Velazquez: Harm Reduction: A Public Health Strategy 216

Part 2: Bias within Science 218
　　Sabrina Rentschler: Unsung Heroes in Science 219
　　Catherine Link: Underrepresented Minorities in STEM 221
　　Claire Lee: Bias in the Process of Research 223
　　Miguel Barranco: Stereotypes in the Scientific World 225
　　Carmen Conroy: Overcoming Barriers in the Scientific Workplace 227
　　Colin Wang: The Importance of Universal Vocabulary 230

Part 3: Taking off the Lab Coat 232
　　Julie Mendoza: Prerequisites for Gaining Public Approval 233
　　Michelle Mendez: Everyone is Included 234
　　Lucero Amaral: Scientific Perceptions 237
　　Gurubala Kotta: How Can Scientists Take Off the Lab Coat and Why is it So Hard To Do So 239
　　Sergio Reyes-Alejo: Taking off the Lab Coat 241
　　Sheelah Bearfoot: Taking Off the Lab Coat: The Mini Documentary 243

Chapter 5: Interpolation of Science 246

Part 1: Intersectionality of Science 248
　　Daniel Russell Cheung 249
　　Michael Ferrin 249
　　Katherin Calero 251
　　Eden Marquez de Leon 252
　　Simon Greenhill 253
　　Matt Kurata 255
　　Chloe Tsang 257

Part 2: Science at the Interface 259
　　Jessica Yescas 260

Osman Shokoor 262
Jeffrey Zhang: The Power and Influence of Popular Science 264
Vin Y. Lay: Science Education in America 265
Pia Choi: Issues with Online PopSci Resources 268
Anonymous: Improving Doctor-Patient Communication 269

Chapter 6: "Class dismissed!" What now? 273

Part 1: Science and Pedagogy 276

Danny Lee: A Look at Challenging the Status Quo in Underserved K-12 Schools 277
Mohamed Taleb: Educacion Vs. Schooling 280
Victor Vargas: "When Will I Even Use This?" 281
Huda Abushanab: Challenging the Status quo in Undeserved K-12 Schools 284
Alexiz Gomez: Step it Up 285
Mike Espino: Teach For American Science 287
Ai Hua 288

Part 2: Future of Science and Sustainability of Change 292

Michelle Guillen: Future of Science and Sustainability of Change 293
Andreas Rodriguez: How are controversial scientific decisions decided on? 294
Grace Ho Jung Kim: Designer Babies: Ethicality 295
Seoiyoung Ahn: The Development of Science and Bioethics 298
Shahil Zhangada: Equipment Used for "Designer Babies" 300
Bryan Huang: Ethical implications of technology in science 302

FINAL THOUGHTS 304

Introduction

By Dax viviD

MCB 15, a course on "Public Understanding of Science," is the only class I know that spent an entire semester critically deconstructing the words of its course title.

To start, who is the "public"?

What are they like? Are they uneducated? Ignorant? Uncontrollable or easily-manipulated masses? Are there common threads that define members of the public? Do they stand in contrast to "private"? One might readily presume, "They pay taxes," but what about children? Or undocumented residents? Of course there is the conception of "the common person walking on the street," who might be considered a member of the public, but what about the people not on the street, like incarcerated individuals? Or people with special medical needs that stay at home or hospitalized? Or people that do not walk but use wheelchairs or other specialized mobility equipment? What about you and me?

Next off, what does it mean to truly "understand"?

How do we evaluate understanding? Can we know if people are just good guessers or nodders? When something is understood, is it held controllable? Does performing well on a test imply understanding? If no test is administered, have we no evidence to claim understanding? What about barriers to language? Just because we might not know the words, can we still understand the concept? Is it even possible to fully understand something, or are there perpetually forming gaps in "understanding" as we change time and contexts?

Finally, we reach the core of the subject matter at hand—"science."

What is science? What is *not* science? Is it a lens to interpreting the world? Does it ascribe causalities probabilistically? Is science universal and completely objective, or is there a subjective, unconscious layer of science that is socially-constructed? Does this layer within the culture of science still count as science? Is science something we can trust? In the interests of what and whom do scientists work? Are scientists a part of the public? Is the public barred from science? What assumptions does each side make about the other? Are there different sides within science? Is science an ever-changing dynamic discourse that shifts with the accrual of unsurmountable evidence? Is science progressive? Is it a series of Kuhnian paradigm shifts?

There are clearly more questions than answers here. While science provides answers over the process of rigorous experimentation and slowly-adapting application, science also creates the conditions of possibility for the emergence of a finding that is unusual and unexpected. This book explores the unusual and unexpected in the public understanding of science.

A couple presuppositions are made of the reader of this book, you. One is that every reader has a personal interest in the topic that is unique from other readers. Another presupposition is that few readers have the luxury of time to read the entire book cover to cover. On the basis of these presuppositions, I hope this introduction will inform readers from different backgrounds on the contents of the book that might be particularly useful to your differing interests. After aggregating essays from students rising from such diverse experiences, we think there is something truly insightful for you in this book.

If you are involved in health work, **Chapter 1: Barriers**, hosts a couple of sections that may spark your interest. Students who have contributed to *How Barriers to Understanding are Barriers to Health, Underserved Communities*, and *Language and Cultural Barriers* share personal experiences and broad overviews of how the infrastructure of health services, specifically in the United States, not only affects the well-being of their respective communities but also how it affects their perception of science. Personal wellness is becoming an aspiration of our generation. There is great value in considering how to positively foster interests in health and influence the perception of the science on which it may (or may not) be founded. One question to ponder while reading: ***what is the goal of health work?*** If it is to cure and prevent, or empower and educate, consider the barriers to these goals as outlined by the students in these sections: Won Jae Jeong, Claudia Campos, Hien Do, Vivian Liu, Brian Woo, Will Lopez, Karen Rodriguez, Tiffany Chang, Reinere Jude Ruiz, Lidia Lisette Pureco, Juan Cano, Alicia Cuevas, Thu-An Tran, Anonymous, Dawa

Gangshar, Glendy Ramon, Joanna Ruelas, Maricarmen Hernandez, Maria Escudero-Fung, Anthony Meza, Claudia Pacheco, and Ruvim Kushnir. Together, these essays foreground minority realities of language barriers and underrepresentation interfering in the process of staying healthy.

Also within Chapter 1, there are a couple sections I would recommend for undergraduate science educators: *The Demand of Non-Scientists to Understand Science* and *On Becoming a Scientist from a Non-Science Background*. While some discussions in undergraduate science education start on the premise that there are science majors and their are non-science majors, I have not heard as much consideration being as explicitly given to students who come from a background that did not emphasize science as a process or approach to understanding the world around us. How can we expect to have students of science from diverse backgrounds if there is the universal, rational, and objective demands of science that align more conveniently (and less antagonistically) with some backgrounds more than others? These sections were thoughtfully organized by the following students: Jennifer Diaz, Claudia Tse, Emani Harris, Khalid Al-Rayess, Xin (Natalie) Yu, Mark Chang, Emily Diaz, Leilani Gutierrez-Palominos, Gila Juarez, Tarun Mendoza, Cristina Batarse, David Lamb, and Siso Phouthavong.

As for K-12 science educators, I would recommend considering sections of **Chapter 2: Public Perspectives**. *Changing Perspectives with Development* reminds us of what excites us about science in school and what affects that excitement. *Science, Family, and Community* opens our eyes to how living in a multi- and transcultural world brings novel insights to science to embrace and support in education, facilitating discussions on cultural humility between teachers, students, families, and communities. These novel insights were brought to us by the following students: Sophie Ballard, Jacob Cota, Betsy Rosales Avalos, Keana Richmond, Diane Chong, Diana Bahena, Vanessa Sarmiento, Mar Jean, Wendi Ruiz, Nancy Cuevas, Rafael Mejia, Adela Ramos, Minh-Thy Nguyen, Maria Rivera, and Marycon Jiro. By integrating these perspectives on education between different cultures and classes, we can collectively inspire methods in teaching and learning media literacy in the face of social media and information overflow at-large.

Journalists are a type of educator; however, if their target audience is "the public," then a lot of assumptions start to be made about who their readers are, potentially overlooking who their readers could be. *Bridging the Gap*, also in Chapter 2, highlights how "the gap" is not simply a discussion of *us versus them* but rather *all of us* as part a heterogeneous mix of resource distribution, determining accessibility to the scientific community and inclusivity therein. Students focusing on the bridge and the gap include Nicholas Garcia, Gabriela Wantah, Anahita Mehrabi, Rattana Sot, Johannah Perez, Paolo Joaquin, Eliott Ahumada, and Andrew Wing.

Additional points for journalists and educators to consider (as well as a scientist, or anyone for that matter, who cares about engaging non-scientists with the joy, beauty, and utility of science) may be found in **Chapter 3: Public Engagement in Science**. Here, two pivotal questions are addressed in the so-entitled sections of this chapter: *"What's in it for me?"* and *How can we engage?* At the forefront of the discussion is the role of education; however, these sections also dabble in elements of politics, performance, and media at-large. What I would emphasize to the readers about this chapter is how "engagement" involves not just dynamic personalities and powerful rhetoric of science "communication" but also learning about the audience from the audience. Communication is an exchange that connects us so we can continuously learn from each other--not a unidirectional transfer of information through a slide deck with a conclusive departure. In these sections, we hope to engage you along with David Barajas, Jaewon (Evelyn) Lee, Miriam Juarez, Elizabeth Aguayo, Austin Weinstein, Liang-Han Tai (Chase), Steven Yu, Josephine Espinoza, Nick Garelis, Deena Abdelhalim, Charles Li, Krystin Ventura, Lorene Cudjoe, Jonathan Homidan, Sarah Cho, and Hung Nguyen.

Whether you are a scientist or a non-scientist, **Chapter 4: Scientists are People, Too: Critical Review of Science** offers a humbling view of the subjective layers of science. In *Ethical Concerns in Science*, we see how the moral dilemmas facing scientists today require a serious look into the history of science, leading us to wonder:

what are we doing today that we will be ashamed of on the basis of ethics in ten, twenty, fifty years from now? One of the approaches to refining our ethics of today is addressed in the section *Bias within Science*. We are all biased, and we are all motivated by certain incentives, whether we like it or not. Instead of pretending to be unbiased and scientistic, with opinions shaped solely by scientific evidence, what if we recognize we are emotionally charged, comfortable with the familiar, and subject to make assumptions of what data implies on the basis of these biases? If this simple introduction does not convince scientists of the importance of self-reflection in recognizing bias, or encourage a follower of scientism to cast a critical eye on what the data "says" (as if data could speak), then a perusal of *Taking off the Lab Coat* may offer you the additional perspectives on the power of subjectivity in science not offered in the summary of a few sentences. The constellation of this chapter, unravelling the threads of the lab coat to reveal the people behind these symbols, is made possible by the contributions of Diego DuBon, Joseline Padilla Alvarez, Joseph Jweinat, Jazmine Carvajal, Anonymous, Terry Kyubin Kyung, Denisse Velazquez, Sabrina Rentschler, Catherine Link, Claire Lee, Miguel Barranco, Carmen Conroy, Colin Wang, Julie Mendoza, Michelle Mendez, Lucero Amaral, Gurubala Kotta, Sergio Reyes-Alejo, and Sheelah Bearfoot. From these essays, we are reminded that an interdisciplinary world can only come from interdisciplinary research. We must work to recognize biases and stereotypes that work against our values and ethics as people.

In **Chapter 5: The Interpolation of Science: How Science Situates Itself in a Larger World**, the network of interactions between what we might call "science" and what we might *not* call "science" is investigated. Whether it be through the *Intersectionality of Science* or *Science at the Interface*, anyone who reads this chapter is offered a glance at how history and culture situate science and how the resulting trickle-effect influences not just the underpinning motivations of the popularization of science but even the interpersonal interactions in educational and hospital settings. The interpolation, intersections, and interfaces of science are interrogated by Daniel Russell Cheung, Michael Ferrin, Katherin Calero, Eden Marquez de Leon, Simon

Greenhill, Matt Kurata, Chloe Tsang, Jessica Yescas, Osman Shokoor, Jeffrey Zhang, Vin Lay, Pia Choi, and Anonymous.

In **Chaper 6: "Class Dismissed!" Now What?** we consider the big picture of what tools are necessary to embrace paradigm changes in science--a chapter from which anyone can gain value and perspective. Along with earning public trust and support by conducting research that directly benefits the people, we wonder how science education can facilitate the ever-evolving dynamic field to produce new, material things while also confronting the social inequities that are bestowed upon us by our predecessors. Contributors to this chapter's vision include Danny Lee, Mohamed Taleb, Victor Vargas, Huda Abushanab, Alexiz Gomez, Mike Espino, Ai Hua, Michelle Guillen, Andreas Rodriguez, Grace Ho Jung Kim, Seoiyoung Ahn, Shahil Zhangada, and Bryan Huang.

What about *Them*? The title of this book is a pronoun we aspire to delete from our daily exchange. In the end, it is truly on all of us to create a multi-directional dialogue that bridges the gap between the ivory tower and the world beyond. Let us claim responsibility to make this change happen. While the expansiveness of change necessary to mitigate social inequalities is lamentable, we can take the first step by not making it *their* problem but **our** problem to solve.

CHAPTER 1

BARRIERS

When traditional methods are ineffective, one must break the barriers and find alternative means of understanding.

Chapter 1: *Barriers*

By Jessica Yescas

Personal health is one of the most important issues everyone has to constantly deal with. Although it may sound unconventional, many are lucky to only have to worry about their health, and not having to worry about understanding the underlying topics related to their health. Many times, the public does not understand health care systems, health conditions, and health risks, not always because they are not fully educated, but because of language, economic, geographical, social, religious and cultural barriers.

In order to define barriers, it is first important to understand the people behind them. Although many people face barriers, it is most common for minorities and immigrants to face barriers because of the start of their integration into our culture. Barriers in this case would be something that keeps people from reaching their full potential. However, as a minority myself, it is difficult to point out these barriers because of how common they are to me. Personally, I have become accustomed to always being a couple steps behind the rest of my peers. I have also become accustomed to not being able to visit a doctor regularly or when I had a health issue because of the lack of good medical clinics around the area or because there just was not enough money for it.

For example, in my culture, it is very common for the doctor to be the last resort when a health issue is brought up. Many times, going to the doctor was the last thought in the minds of parents and so home remedies were the answer and solution to common colds and fevers. Though these home remedies might have been an easy, cheap, and quick fix at the time, it never occurred to them that these small colds could be an underlying symptom of something greater. There would never have been a way to know unless one had had a background in a medical profession. Through these home remedy practices also comes the mistrust of medical professionals, which demonstrates the consequences of being embedded into a culture so deeply, one starts to take customs for granted and not even question the ideology behind them and how they could potentially be social and cultural barriers.

We would all like to believe that America is no longer the melting pot that it used to be defined as, but rather more of a mixture of various different cultures; however, there still exists some barriers that force diversity to wither. In healthcare specifically, the most prominent and perhaps the most important of these barriers would be language. These language barriers including but not limited to the differences in medical jargon and the differences of languages between the patients and doctors.

This common barrier that many immigrants come to find upon going to the doctor is the language barrier. *Imagínese que usted va al médico porque usted tiene una pregunta sobre su salud. Cuando finalmente le llaman para entrar en la sala con el médico, usted empieza a describir en detalle su pregunta. Pero para su sorpresa, él sólo le da una mirada confusa y dice* "I'm sorry, I don't speak Spanish. Could you say that in English?"

Barriers also interfere in the lives of the people who are in need of learning more about their health and those who desire to know about the health sciences. For many people interested in learning more about their health and ways to improve it, it is often difficult to do so if coming from a low socioeconomic background. This is true especially if one is from a geographical location consisting of mainly minorities. People stuck in these situations--in which they are not able to keep up with scientific info and are restrained from becoming involved with the scientific world--are unfortunately very common.

What if you had always wanted to learn more about the sciences and you thought of going to medical school, but you believe that your education is not as up-to-par as other students? Well, just like all other underserved communities including first generation students, these are common occurrences when growing up in these communities. However, what really is considered science? To many, science means working in a lab; but the true form of science is the application of the scientific method. So, in reality, we all tend to apply this method in our everyday lives, making us all scientists in some type of way. Nevertheless, because of this recurring stereotype of a scientist, students in specific often feel pressure to conform to the science that society thinks they do. Many times these students do not know true science because neither they nor their parents were allowed to opportunity to learn about the sciences.

In the following chapter, these any many other similar situations will be recounted through personal experience or interviews in order to demonstrate the need for the public--and those in the sciences--to understand the need for a bridge for the gap between them by first breaking down these barriers.

Part 1: How Barriers to Understanding are Barriers to Health

We aim to recognize many different types of barriers, e.g. language, cultural, knowledge, access barriers, and understand how they interconnect and work together to form a bigger barrier to health.

Won Jae Jeong: What Does That Even Mean?

With rapidly developing field of molecular biology and medicine, there are overflow of information that is impossible to digest all. For example, with the development of the field of genetics and genomics, we will live in a world where treatment of disease through gene therapy is imminent. However, the title of this paper, What Does *That* Even Mean?, is my interpretation on the public's impression when they first encounter cutting-edge scientific information. In few interview conducted on genome editing technology called CRISPR, the public could hardly understand the science behind many technologies and questioned: What is DNA? And *what does it even mean* to modify my DNA?

Upon collecting responses of the public about genetic modification, I found a disturbing clip from Youtube, named "Genetically Modified Organism (GMO) - Myths and Truths" (link attached below). The video, showing as if it is from a published article, narrates "it is a laboratory based technique where a foreign gene is inserted into the DNA of the plant. This is an uncontrolled process, because the site of insertion is random and may potentially damage the plant's genetic makeup". Such claim is very fallacious because there are many genetic tools that allow one to screen for unwanted mutations and to target specific sites of genetic alteration, such FLP/FRT, Cre/Lox, and CRISPR, etc. Knowing that this claim is not true from knowledge and experiences working in a genetic lab, I looked for the source of it in Pubmed. Unsurprisingly, such information was not in any published article.

Through the Open Access Movement which releases many published journal articles for free, the public is now granted with access to many published articles; but do they? It is very challenging for the public to understand technical terms and process the information without any background knowledge in the field. Also, because they lack technical knowledge, few wordings on the paper or social media may mislead and lead them to misunderstand. The public cannot make informed decision about their health when they are not informed with the knowledge.

Genetic modification may not be the most worrisome problem, but aforementioned cases could lead to more endemic problem: such misleading claims that insist and promotes suspicion on scientific researches possibly increase the public's distrust toward the science in general. For example, falsely believing that GMO is a randomized creature that is detrimental to one's health and thus growing skepticism toward the science, one may start to doubt the science behind vaccination. Searching through social media, one may find that vaccination causes polio and deny to vaccinate. Although there are tremendous information that supports for vaccination, if one is skeptical, even a single information against vaccination can be propagandizing.

It is impossible to destroy the barrier by targeting a specific barrier. Upon dealing with the barrier, it is not difficult to realize that the barrier is indeed consists of many other barriers that are interconnected. If one resolves the issue of access to information through the Open Access Movement, there emerges the problem of misunderstanding the information by the public. If one solves the problem of misunderstanding the information, there emerges other barriers such as language barrier that one simply cannot understand the language, cultural barrier that cultural tradition is against, for example, unnatural modification, religious barrier that claims the science is "un-God-ly", involvement barrier that one is imply indifferent, etc.

Because these barriers are highly interconnected, more holistic approaches should be taken primarily to minimize the gap between the scientist and the public before any information is conveyed from the scientist to the public. Scientific community should not consider themselves as superior to the public and assume the public cannot process the science; the public should stop doubting the science due to its difficulty and be more willing to be educated to make informed decision. Such general attitudes toward each community, granted it will take long time, can be improved by, for example, raising general education level and awareness campaigns. Once this primary barrier between the public and scientists is resolved, it would be more facilitating to build the bridges over many barriers to achieve better health of our communities.

Link to the Youtube Clip:

https://www.youtube.com/watch?v=M_ztZGbLEJ0

Link to the Group's Video Presentation:

http://mcbfifteen.tumblr.com/post/118553156485/barriers-how-barriers-to-understanding-are

Claudia Campos: In Need of Better Distribution of Knowledge

When many of us hear the word "health", the two associations that cross our minds are maintaining a healthy diet and exercising. The way it is advertised makes it look like an easy thing to do, especially since the media has access to the perfect settings and props to make everything look ideal. In reality, how many people live lives like those being advertised on commercials or in magazines? Not too many. This is one huge issue within our society and being effective in the process of communicating health related science to the public, because advertisements usually target one group, when in real life there are millions of different individuals. That means that simply because one method works for certain groups of people does not mean it will have positive impacts on all.

This is important to highlight because it influences decisions to where advice for healthier lifestyles is being obtained. For example, I have witnessed colleagues invest money in magazines such as, "HEALTH," "FITNESS" and "SHAPE," where they seek expert advice on exercises and diets. I have asked why they seek information from these sources and a common answer is that they are the most accessible and affordable things to them. The most interesting magazine I once saw a colleague purchase was one on diabetes, because they were told there was a possibility they could have high blood glucose.

Terms such as diabetes, high cholesterol, and high blood pressure, are becoming more common in regards to impacts on individuals health. Yet, many people can only sit back and question why it has happened to them. The unfortunate part about people reaching this stage in their life, when all they can do is ask how or why, is that for many of these chronic diseases there are answers (science behind the reason why). Possibly, if the kind of information was distributed more effectively to communities who have higher chances of developing any kind of chronic disease, early precaution measure could have been taken into account. By doing so, we would be looking into increasing their chances of changing ways in their lifestyles to reduce their chances of developing a disease as such.

According to a presentation given by Ahna Suleiman, a postdoctoral fellow working with Ron Dahl in the School of Public Health at UC Berkeley, the top 'Ten Leading Risk Factors for Preventable Disease" include maternal and child underweight, unsafe sex, high blood pressure, tobacco, alcohol, unsafe water / poor sanitation / hygiene, high cholesterol, indoor smoke from solid fuels, iron deficiency, high body mass index or overweight. She mentioned how these ten leading risk factors are all connected to behavioral decisions. In addition, she touched on the fact that society tends to blame individuals for being affected by these risk factors, when in reality; the environment has large impact on why these diseases develop.

It is very difficult to find one solution in order to bridge this gap, and I am aware in can take years for anyone to possibly find a solution, but it is key to be able to have these conversations on barriers between the public and science to raise awareness about the kind of information that exists. I believe that by distributing flyers, brochures, and by having more community health fairs, we can truly spread more science facts to communities that are at higher risks for developing chronic diseases. We should work harder to break the stereotype of saying things like, "You consume too much sugar, that's why you got diabetes," and pay more attention to the outside factors contributing to one's health.

Bibliography

Suleiman, A. (2015). Health and Social Behavior An Overview [PowerPoint slides.]
   Retrieved from
   https://bcourses.berkeley.edu/courses/1303119/files/folder/Lectures?preview=5
   6465725

Hien Do: How Barriers to Understanding are Barriers in Health

With new policies, innovations, and changes in medicine occurring daily, it becomes increasingly difficult for the public —who I believe to be people who are not involved in, invested in, or immersed in health, medicine, technology, or biological or physical sciences– to have a strong grasp on staying up to date, understanding (being able to critically analyze) what is happening to their health and healthcare. As my group and I explore barriers in attaining higher or formal education, in learning technical subject-specific terminology, in accessing resources such as scholarly journals, and even in being exposed to the field of science, I see how all these barriers interconnect and play a role in how the public internalize information to make decisions about their health.

Unfortunately this gap in understanding creates miscommunication that I witness frequently with my parents, who do not hold college degrees, are limited English speakers, and live in an underserved community. They often are misinformed or uninformed on scientific topics that relate to their health. This ranges from political science of policies or acts that affect them, research on preventative and therapeutic medicine, diet, exercise, etc. Due to my parents' lack of formal education of technical terminology, poor English proficiency, and lack of dissemination on the sides of research and policy, my parents often get their information from other members of the community with similar socioeconomic conditions or hear bits and pieces from the radio or television –which are often quite biased.

Recently my parents were shocked about an additional section they had to complete for their tax form. The form requested information about their health insurance information; this is in accordance to the Affordable Care Act's mandate for personal health insurance and those that fail to comply –either having unacceptable or no health insurance– will be fined. As I explained to them sections of the ACA that would affect them, they repeatedly told me that no one had explained the Act to them, and that they did not understand any of the information they did receive, such as pamphlets or short notes that their doctor's secretary basically gave them to leave with, without any explanation. They also did not understand a letter they received informing them that due to the ACA's effect, their Medi-Cal benefits became more comprehensive. Learning this, my parents were encouraged to schedule a check-up they had been putting off for a long time due to financial fears. They were surprised when they learned the ACA gave them a lower deductible and better coverage, since they had heard on and trusted a relatively conservative Vietnamese radio station that denoted ObamaCare as an expensive economic and government cost that would take away autonomy from US citizens. In addition, my parents also ask me about changing their diet or lifestyles based on articles and news segments that denote causality – when the research only concludes correlation. An example would be a study stating a potential correlation between drinking "moderate" coffee and lower risk of heart conditions, where the news station failed to mention that it was a cross-sectional observational study done in South Korea, in which it is difficult to extrapolate data from since this study only occurred during a relatively short period of time, is subject to human biases, and takes place in another country with different lifestyles, diet, and habits. However my mother was convinced in justifying her morning coffee and considered drinking more, as the news never mention the potential risks of coffee such as caffeine addiction or sugar intake, and disregarded giving a clear definition of "moderate" coffee consumption.

Drawing from my parents' experiences I see a relationship between understanding and health, and I hope to improve this especially in vulnerable communities by looking at how to properly disseminate information to the public. There are many things to consider since both the information and the public are widespread. Is it that information is being held back? Has and if so why has a distrust in medicine and research formed? Or is there too much trust in medicine and research (potentially as a result of deferment to authority figures)? Is and how is technical terminology and language turning people off from science? How do we convey information without this terminology, but maintaining important details? How can we help facilitate access to resources?

I recognize that it is not profitable to give attention to underserved communities, as it is additional work policy writers, researchers, the medical field, and more, have to take; however, I believe this extra step towards these communities result in greater returns than money. This bridging of the gap can support and aid many people, including the next generation, by better supporting their health and lifestyles, allowing us to progress as a healthy and advancing society. That is why I hope for effort on both sides, where policy makers, researchers, journalists, media, and health professionals to work together and actively engage with the public to properly disseminate their information without bias, allowing the public to make informed decisions. In addition I think that if possible, the public should educate themselves using accessible or available resources, the Internet, libraries, databases, and more, so that they can question and critique the information they are being given to help them make changes or maintain their lifestyles to improve their health. I see that there is community level work being done by some organizations (in my community, these consist of free or affordable clinics, health fairs, and patient advocacy services), with the creation of short pamphlets, leaflets, and notes that use less technical terms or disseminate information that can be passed out and around on a wide scale. This is a great solution to me but they are often lacking finances, workers, and structure, so I think that they should be granted government subsidies or better privately funded.

I would like to re-express that this write-up is based on my opinion and personal experiences. I also want to address that defining who the public is, what is science, and what understanding means is very difficult. There are no clear binaries as there are many overlaps, gaps, and technicalities, and this ambiguity seems to results in different definitions of the terms based on personal experience and background, so I assume that my definitions will differ from the readers'. But I am in no way trying to generalize or speak for others; I merely want to start a conversation on addressing this gap as it results in a direct effect on my family and my community. By giving my personal insight, I hope that it triggers a spark or adds meaning to others' perspectives on the matter, so that hopefully the readers, others, and myself collectively recognize this existing problem and collaborate to work towards a solution.

This is a link to a video my group mates and I made to share our personal opinions on the topic. We all have a different background and hope to reach out to others who can connect with our experiences and for those who do not, to see that there is a pattern worth looking into.

Michael Ma: Cultural, Social, and Political Tides of Misunderstanding as a Threat to Health

Video Link: http://mcbfifteen.tumblr.com/post/118553156485/barriers-how-barriers-to-understanding-are

Recent health crises have reflected the perpetuation of obstacles to public health as a result of public misunderstandings. Despite advances in science and proven facts that have been established and disseminated to the public, there seems to be a perpetual resistance to adopting the recommendations of scientists - a general tide of distrust. At the root of misconstructions and refusals to change are cultural, social, and political circumstances that produce barriers that have tangible detrimental impact on public health. Perhaps most conspicuous of this disconnect is the recent vaccination controversy that has been connected with a severe outbreak of measles in the United States earlier this year.

This public health crisis reflects misinterpretations of the general public that have obstructed the administration of a vaccine that has been proven to be scientifically sound. This is certainly a public health concern, especially for members of society who are immunologically deficient and cannot safely acquire the vaccination. Specifically, there has been a proliferation of disreputable information online that continues to assert debunked scientific facts, such as the claim that vaccinations promote childhood developmental disorders like autism. While there are scientifically accurate resources that are widely available in the form of pamphlets or online articles, many are unable to access these educational resources because of barriers such as language. Many resources are only printed or available in major languages such as English and Spanish, but are not available in minority languages.

Another major obstacle that leads to misunderstanding in the context of public health crises is the business politics of the healthcare system. The current health system is faulty in that doctors are not adequately compensated for spending the time to educate their patients on issues such as vaccinations. Because doctors are not appropriately compensated for their contributions to encouraging the public, they often do not carry out the task of informing patients about the facts of vaccination and disease.

Furthermore, the media propagates fears by publicizing confusions over science, muddying the waters of scientific fact and frenzying the tides of confusion. News media organizations often take advantage of controversies in healthcare to bolster ratings and viewership, producing sensations of fear where there needn't be any with the backing of science. Dr. Gary Procop, a fellow the College of American Pathologists, stresses in regards to parents who choose not to have their children vaccinated, "In many instances Americans will listen to television personalities rather than scientific opinion. As medical professionals we have to be vigilant, diagnose accurately and collaborate with out public health colleagues" (Science Daily 1). Therefore, public health professionals face the challenge of overcoming misrepresentations of vaccinations in the face of popular media and misleading sources of information.

Perhaps of most concern to the general populace is that many people lack the understanding of the ramifications of choosing deliberately to not vaccinate themselves and their family members. I was afforded the opportunity to speak with Professor John Swartzberg, a clinical professor in UC Berkeley's School of Public Health. Swartzberg stressed that the recent measles outbreak is the outcome of people who do not adequately comprehend that their choices affect others in the community. Specifically, when people choose not to vaccinate, they are leaving at-risk populations vulnerable to suffering the effects of the disease. In particular, young children who are not old enough to get vaccinated, as well as immune-compromised people who suffer from diseases such as AIDS cannot get immunized by vaccines; therefore, they are dependent on the rest of the population to be responsible and educated when considering vaccinations.

To resolve the disconnect between science and the general population, there are a variety of approaches that should be considered. The health system may benefit from reformations so that doctors are compensated for their time educating their patients about the importance of health procedures, including vaccination. Health professionals need to be able to communicate to their communities accurate information and provide resources to aid families in making decisions that have a tangible impact on public health.

Political conflicts have emerged as the culmination of barriers to the comprehension of public health, with its social, political, and cultural complexities. Will the imposition of new laws serve as a sound resolution to misunderstandings, or will they only fuel the flame of opposition? Even government leaders and political institutions seem to be afflicted in controversy over misconstructions concerning vaccinations. All in all, complex factors account for the vast ocean of public health crises that have arisen and are likely to arise due to miscommunications in science understanding.

Works Cited

College of American Pathologists (CAP). "'Herd immunity' threatened in measles outbreak, pathologist says." ScienceDaily. ScienceDaily, 5 May 2015. <www.sciencedaily.com/releases/2015/05/150505182610.htm>.

Vivian Liu: Language Barriers to Understanding Health

Everyday, there are immigrants entering and leaving the United States. This opens up the doors to new languages, creating a mixture. Unfortunately, this poses a major problem in health care system. Without a commonly understood language by both parties, it is difficult for doctors to communicate with immigrant patients, thus creating the barrier for patients to comprehend their health. This is connected to our group's overall objective because we all see different barriers to health-- one of them being lack of access to scientific knowledge. The language barrier that I am focusing on causes people to not comprehend what they are being told so they do not have access to that certain knowledge. This may discourage some of them from seeking help from health workers due to the language barrier.

Before I was born, my parents emigrated from China to California. When I was a little girl, my Cantonese-speaking mother would take me to Asian Health Services in Chinatown, Oakland. Despite being located in Chinatown, my doctor did not know how to speak Cantonese so we often requested for a translator. We are lucky that we speak one of the more well-known languages, but for those who speak a less well-known language or have a smaller population size in the community, they were not fortunate enough to have translators. However, even with the assistance of third-culture translators, there is still a loss of information because they may not know all the medical terms in one of the languages. As the years elapsed, my relatives from China also immigrated to California and encountered the same language conflicts as my family. One of the consequences is that my relatives discontinued visiting the doctors who practice Western medicine, and instead, they relied heavily on alternative Chinese medicine. However, currently, health centers are implementing programs that provide alternative medicinal practices.

During the summer of junior year in high school, I had the opportunity to intern at Frank Kiang Medical Center in Oakland where they target underserved, minority communities. While I was working in the Vital Signs room, the two nurses there spoke at least two different languages fluently (not including English). A lot of their patients did not speak English, so the nurses were extremely vital in communicating with them. When I shadowed Dr. Suzanne Nguyen, she informed me that her Vietnamese skills are very limited and basic. Nonetheless, she is still able to communicate with her patients daily but they may lack access to the full understanding of their health due to poor communication. The same goes for the other bilingual doctors who work there. If we had more health workers who spoke fluently, then that could potentially bridge the gap between language barriers and health.

Other solutions to fix this problem include organizations that translate for patients and doctors. For example, on UC Berkeley campus, there is Volunteer Health Interpreters Organization and Labor Coach that serve to translate doctor-patient conversations and documents in many different languages. Labor Coach provides services for pregnant minority woman who do not speak English. The students are in contact with these women throughout their pregnancy until childbirth. When they are in labor, the students go to Alta Bates in Berkeley to assist them emotionally and facilitate discussion between doctor and the women, explaining what is happening and their current condition. When I applied for this program and was interviewed, I struggled a lot of with translating certain words simply because they were medical terms that were never spoke of in my house. It is hard to imagine what it is like for these women to not understand what the doctor is talking about without translators. Thus I am grateful that these programs exist to help non-English speakers. Nevertheless, the gap between language and health is still growing with the influx of new immigrants everyday.

Link: http://mcbfifteen.tumblr.com/post/118553156485/barriers-how-barriers-to-understanding-are

Brian Woo: Personal Evaluation of Religion and Science

As a young child growing up in a conservative Christian community, my experiences with science, and evolution, were mixed. I remember the first couple of books that I picked out for myself at the library were on viruses and cancer, and it was through reading those books that I realized I enjoyed learning about science, that science was *cool*; in fact, had I not had that early exposure to science, I cannot say whether or not I would be studying biology right now in college. At the same time, however, some of my church's members (and more importantly, teachers) taught me in Sunday school (a teaching session held after sermons) that science—specifically, evolution—was flawed.

As I grew older, I began to question and analyze this sort of dichotomy between conservative Christianity and science. I learned about evolution in my high school biology courses and also through own personal reading of online articles and books; at the same time, I continued to attend church and talk to my fellow church members about their stances on evolution and science as a whole. I began to have a good idea of what some of the arguments were on 'both sides' (i.e. I learned about the idea of intelligent design and others like it in church, and also had a good understanding of the mechanisms of evolution and evidence for it). I learned how ugly arguments against evolution could turn, and how impassionately 'the public'—or in this case, people in my religious community—could argue against evolution, even when using completely wrong information and putting up straw men; for example, I quite vividly remember hearing someone in my church say one time about evolution, "Evolution doesn't make any sense. If it were true, then why are there still apes?".

However, even though I continue to stay true to my faith, I do not see any necessary gap or schism between the two (religion and science). Admittedly, throughout my high school years, I was annoyed by the ignorance of evolution that I saw displayed both throughout high school, in my religious community, and on the internet; even in my advanced placement biology class, there were still at least two or three people that I remember talking to about evolution, and whom I remember blatantly refuting evolution and calling it unsupported, or improbable (this was during the evolution unit itself). I have always seen the theory of evolution as something very logical and intuitive, and the fact that it is so well-supported and articulated only serves to reinforce my understanding of it.

From taking this course, I realize just how important it is to 'bridge' this gap. Through being misinformed about evolution, people like those I encountered in church continue to despise science and misinform others; they vote using misguided information about science, and come to disregard science as a whole. I realize that even though a disbelief in evolution may not directly be related to issues in health, I strongly believe that this disbelief extends to science as a whole and has major negative impacts on issues that *are* health-related, e.g. vaccinations. To be honest, I am not sure exactly how to go about remedying this gap; I can only speak from experience that it causes otherwise reasonable people to make completely wrong statements.

Will Lopez: Education

For our final project my group focused on the many different barriers that people face when it comes to understanding health and science. Many of the barriers society face range anywhere from economic, religious, social and educational, amongst others. In my write up I will focus on one of the barriers that I experienced growing up, which is the barrier of education. Growing up I attended an elementary and middle school in which academics were not very strong. Since most of the students in my school came from low socioeconomic backgrounds, the standard of education was never as good as other schools, which received more funding. In the area I grew up most parents did not go to college and therefore never received a formal education. In addition my mom along with my friends parents were immigrants from other countries and they had a very difficult time speaking English. I remember always going to the doctor with my mom and she would constantly ask me to translate because she had no clue what the doctor was saying. As an eight-year-old child it was tough for me having to translate everything to my mom, especially since I also did not know what the doctor was talking about a lot of times. I clearly remember that elementary and middle school were very laidback and I would never spend time doing homework let alone studying. I was never interested in learning because I never had that push from others to work hard. When I went to high school I moved to Irvine, a city where the median income was six figures and 70% of adult residents had completed a bachelor's degree. I remember my freshman year was very tough since I had never been exposed to such a rigorous course load, luckily my teachers were very invested in helping me succeed and I would go to them for extra tutoring after school; something that I never had access to in middle school. Eventually I started to copy my peers study habits and for the first time in my life I felt proud of myself, I was learning more than just the basics. I feel like my personal story epitomizes how our educational system is flawed and this in turn allows some children an advantage over others when it comes to understanding science and health. Looking back, most of my friends from middle school did not end up going to college and therefore never received a great education, however all my friends from high school ended up going to four year universities, such as Harvard, Princeton and UCLA. I think that nowadays many people fail to realize that the there is a direct correlation between the education you receive growing up and your knowledge of the world in general. This includes the public understanding of science and health. I think we could bridge the gap that exists in understanding health by first reforming the education system. We need to equal out the playing field by first providing all students the same opportunity to learn and thrive. I think it is crucial that we start implementing an interest in science from a young age, since this is the time when a child is curious about everything. Although not everyone grew up the way I did and I do not want to generalize by any means, but I strongly believe that there is a cycle in which educated people live better lives and therefore are able to provide their children with a better education. On the other hand people who never received an education are less inclined to push their children to learn and work hard in school so their children never develop a desire to widen their intellectual scope.

Part 2: Underserved Communities

The goal of our group will be to focus on the lack of resources, which can take the shape of education, food, transportation, health care, or professionals, found in underserved communities and explore possible solutions to mitigate the adverse effects of scarcity in these marginalized populations.

Karen Rodriguez: The Lack of Access to Community Resources In Underserved Communities

The inaccessibility to community resources in underserved communities is one of many factors that display the disparities present around the world today. These disparities are the number one leading cause of the current gap present among communities composed of individuals from different socioeconomic status. Although these differences are one of the many causes of the unequal distribution of resources such as healthcare, amenities and education, the lack of access to fresh, healthy and affordable food continues to be one of the biggest problems.

Although access to proper nutrition is a topic of extreme importance and something we should pay close attention to, it seems as though it is something that a lot of communities lack. According to the United States Department of Agriculture, approximately 2.3 million people (nearly 2.2% of all US households) live in low-income neighborhoods that are located more than ten miles away from a supermarket. Since living conditions have been scientifically proven to impact every aspect of individuals' lives, it is important for us to focus on equally distributing resources in communities around the world. However, that might be easier said than done and although there are communities with access to a local grocery store, this does not necessarily mean that healthy, nutritious food is actually accessible to community members; whether it is due to cost, preference and/or variety [or the lack of].

The lack of nutritious healthy food does not only come from the inaccessibility to local groceries stores. Therefore, because our overall health can be seen as a continuum, the problem should be targeted at multiple levels in order to initiate change. Schools can be said to be a very important level in the sense that in the U.S, state laws require children ages six to eighteen (on average) to attend school. That being said, because children spend a significant amount of time on school ground, it is important for them to have healthy food readily available. However, for public school districts like San Francisco Unified, this can be challenging simply due to the lack of funds. According to Professor Kristine Madsen's, an environmental approach to creating an effective change within the public school system is one if not the most important simply because the way in which food is set up in the cafeteria truly affects children's food choices. Often because lunch time is only thirty to forty minutes long, kids make food choices based on what is easier to reach simply because as Professor Madsen mentioned, they try to maximize their time in order to spend time with friends. Often, what is easier to reach is not always the healthy nutritious food options therefore this is also something we should definitely examine closely.

That being said, in order to effectively target the food environment, one must think of sustainable policies that will work not only in underprivileged communities but also within the school environments located within these communities. According to Gibbons, in her journal *Marketing Healthy Choices in the School Cafeteria*, pricing, placement, visibility, promotion, marketing and variation in foods are all things that can help create simple, sustainable changes to kids' food choices on school ground. Additionally, for individuals from underserved communities who lack a local grocery store or who cannot afford to buy healthy, nutritious food at their local super market, perhaps growing fruits and vegetables in their back yard or if this is not an option, possibly even a communal garden can allow individuals gain access to healthier choices. That being said, because the disparities present in underserved communities are endless, making small sustainable changes like those listed above, can create positive long-term effects on communities around the world.

Group's Media: https://youtu.be/RpapxCxs_kA

References

Gibbons, Heather W. *Marketing Healthy Choices in the School Cafeteria* (2009): 3-43.

Jan. 2009. Web. 3 May 2015.

National Conference of State Legislature. 1 June 2010. Web. 10 May 2015.

United States Department of Agriculture Economic Research Service. "Access to Affordable and Nutritious Food: Measuring and Understanding Food Deserts and Their Consequences." United States Department of Agriculture, 2009. Web Accessed February 23, 2015.

Tiffany Chang: Access to Nutrition and Education

Access to food itself may not be that much of a challenge, but access to the right kind of food may be for some. Now when the lack of access to the right food is combined with barriers such as socio-economical and educational limits, the societal issue becomes even more serious. According to the Centers for Disease Control and Prevention (CDC), at least a third of the American population is obese. The ailments don't stop there since obesity is linked to a number of other common conditions such as diabetes, heart disease, and stroke. All of the conditions listed are preventable, but prevention becomes difficult when an individual doesn't have the time nor the financial means to prepare a home-cooked meal as well nor the awareness that certain types of food are harmful to human health. These obstacles are especially prevalent in underserved communities. The existing cultural and language barriers disallow information to be transmitted, so many in the undeserved communities go along unaware.

Due to this setback, it would be great to see more opportunities provided in the community, in a variety of languages, whether in the form of classes or fairs, so individuals can become more alert of healthier options and the consequences of eating unhealthy. Also, emphasizing these topics in the class curriculum could be beneficial since students can take control of what they consume and hopefully be able to share the information with their friends and family. The Berkeley Unified School District has already put this plan into action by having school gardens so students can grow their food and learn about healthy food choices (The Edible Schoolyard Project). So far, the new curriculum successfully engages students to think about which foods are good for the body and even how to prepare the meal. The students can share their experiences with friends and family at school-wide Family Night Out events, at which knowledge and encouragement for eating healthier reaches the community. The idea of implementing this activity into the school curriculum seems to be a good way to get students excited about leading a healthy lifestyle. Now if this curriculum could be introduced to schools in other communities, the students would be able to bring the information back home to their families, and we could move one step closer to spreading awareness about eating healthy.

Aside from introducing the information to students, information can also be spread through community fairs and events. From personal experience, just by being able to talk with members of the Berkeley community as a part of D.U.L.CE, the diabetes awareness organization on campus, I get the opportunity to explain information such as how to read food labels and what proportions of each food group make a healthy meal. One example is the sugar demos that we have on display—so there would be a soda can with the amount of sugar that is in the drink in a plastic bag next to it. This attracts attention, and parents would frequently come up to the booths asking for information, and it is rewarding being able to provide information that they may not have heard before. Usually, our table is one of many that are health-related at these events, so individuals can visit and gain information from other tables as well. As a plus, our team is multilingual so in case a member of the community doesn't speak or understand English, someone on the D.U.L.C.E. team can explain in the needed language. Pamphlets are also available in both English and Spanish in case individuals do not have time to stop for a long time at the table. Of course, the information isn't meant to be forced onto the community as there are cultures to respect, but hopefully small changes in lifestyle to overall improve or maintain health can be taken into consideration. By these methods, the awareness gap can gradually become minimized, and as a result, the lack of access to nutrition may also be minimized.

Resources:

"Adult Obesity Facts." Centers for Disease Control and Prevention. Centers for
    Disease Control and Prevention. 09 Sept. 2014. Web. 25. Apr. 2015.

"The Edible Schoolyard Berkeley." The Edible Schoolyard Project. Web. 7. May.
    2015.

Reinere Jude Ruiz: Underrepresentation in Medicine

After listening to Professor John Matsui's lecture on the significance of cultural humility and diversity in the sciences, I began to critically think about the role of minority physicians in alleviating health disparities in underserved communities. In the research article "Cultural humility: Measuring openness to culturally diverse clients," authors Hook, Davis, Owen, Worthington and Utsey conceptualize cultural humility as the "ability to maintain an interpersonal stance that is other-oriented (or open to the other) in relation to aspects of cultural identity that are most important to the person" (2). In other words, physicians and other health professionals who exercise cultural humility are continually learning, adapting, and appreciating the culture of their patients. They develop a respectful partnership with the people they treat and they provide healthcare that focus on a client's cultural needs.

Minority physicians, in particular, can empathize with low-income, disadvantaged, and marginalized populations. Unfortunately, however, the current physician workforce is not reflective of the United States population: "Racial and ethnic minorities comprise 26% of the total population of the United States, yet only roughly 6% of practicing physicians are Latino, African American and Native American" (AMSA 1). This is because pre-health minority students historically struggle financially and academically. There are institutional barriers for these students, and they are not likely to reach their career aspirations. Underserved individuals consequently lack access to physicians who personally relate to their troubles and their experiences. Thus, to provide more culturally appropriate healthcare to disadvantaged communities, we must increase the number of underrepresented minority students in medicine.

To help achieve this, I believe more support programs, funding, and scholarships must be dedicated to underrepresented minority students who are in pursuit of careers in the sciences. For example, initiatives such as the Biology Scholars Program and the Summer Medical and Dental Education Program both aim to increase the number of highly qualified science and health students from non-traditional groups. They provide a support system and a sense of community to students who lack the resources and the opportunities to succeed. These programs are also present in the elementary, middle, and high school levels. Several of my peers have participated in Upward Bound at the University of California, Berkeley and

SMASH Academy at Stanford University, which are programs that promote science, technology, engineering, and math (STEM) enrichment in underserved communities. These organizations serve as pipelines for minority students and act as vehicles for progressive change in the medical system.

All in all, as a low-income, Pilipino-American, first-generation college student, I personally resonate with the subject of underrepresentation in medicine and barriers to the sciences. There is an evident need for culturally humble and diverse physicians in underserved communities, and luckily, more and more programs are materializing to address this problem.

Bibliography

American Medical Student Association. "Enriching Medicine Through Diversity." Web. 25 Apr. 2015.

Hook, J. N., Davis, D. E., Owen, J., Worthington Jr., E. L., & Utsey, S. O. (2013). Cultural humility: Measuring openness to culturally diverse clients. Journal of Counseling Psychology. doi:10.1037/a0032595

Lidia Lisette Pureco: Social Workers

Since I was young, I have witnessed first-hand the barriers that can arise from living in an underserved community. Underserved communities, which can be defined as specific areas or populations that lack quality services, ultimately arise because there isn't the sufficient resources available to specific communities. I, however, consider the problem to be rooted beyond the lack of resources and consider that the way resources are distributed also plays a role in the creation of underserved communities.

Growing up in a small town where the majority of the families where low-income, I witnessed the lack of healthcare not only within my own family, but throughout my entire community. This pattern transcends throughout low-income, minority families in which they never find the appropriate resources to their issues because the process can be "too complicated". When I was younger, my parents spent many years without medi-cal because it meant having to miss work and spend hours waiting for a social worker to help them fill out the paper work. Since they didn't have the proper insurance they could not visit the doctor because they couldn't pay out of pocket.

The reality of this situation is that it has nothing to do with the social workers; instead the problem is rooted with how the work is shared. A social worker named Sherri said, "I'm an advocate for people who don't know how to go through the system. I help people break through the obstacles that could keep them from getting things done and help them negotiate to get what they need" (qtd. in Edison) which truly encompasses their role in the patient's life. However, there are many obstacles that interfere with the quality of service provided. While some situations may be more difficult than others, many social workers often find themselves feeling overworked which can result in not finding the necessary resources or even overlooking a patient. Just in November 2013, Los Angeles County social workers filed a lawsuit demanding smaller caseloads and hiring 1,400 social workers. These demands, they claimed, would help them provide better help to their patients ("DCFS").

In addition to hiring more social workers, I believe there are also better ways of providing patients with the adequate resources. An example of a program that is designed to help address the many social, economic, and legal issues that patients may be is facing is the Highland Hospital Advocates, which is based in Oakland's Highland Hospital. Ultimately the goal of the program is to provide the patients with resources with the partnership of the social workers. I have personally worker with the program and have witnessed firsthand the positive outcomes. Working with this low-income population, but from more of a social worker aspect, I have realized that it's not only being able to provide the person with the resources, but finding a way to adequately provide that resource by learning more about the patient on a one-to-one basis. As California, we are a melting pot of different cultures and it's only correct that we all respond to situations differently. For that reason, I see the individuals providing these resources as an important tool to overcoming the barriers within underserved communities.

Works Cited

Edison, Ilaina. "Social Workers: Giving Voice, Navigating the System, Reclaiming Lives." The
Huffington Post. TheHuffingtonPost.com, n.d. Web. 6 May 2015.
"DCFS Social Workers Demand Smaller Case Loads." ABC7 Los Angeles. N.p., n.d. Web. 27
April 2015.

Juan Cano: Como Geografia Es Un Factor Important Que Excluye Comunidades Subatendidas De Recursos Basicos

Historicamente hay eventos que repiten; por ejemplo cuando comunidades son empujadas lejos de recursos basicos hay potencia para humanizar y segregar. El transporte y la seguridad son algunas de la barreras al acceso a la información científica a través de servicios de salud e educación como las escuelas e biliotecas. Esto ocurre *en las comunidades* como Richmond, CA e valle central donde ahora continua excluir comunidades subatendidas de accessar cuidado de la salud, educación, transportación, etc. Quiero explorar las cuestiones que avecinan por la geografica localidad de comunidades subatendidas. Por ejemplo, el medio de transportación en la area de la bahia es mucho más accessible a comparado por las personas que viven en el valle central lo cual es mas dificil accede a recursos como información, talleres profesionales, sitios de comida, cuidado de salud, e educación. Adicionalmente, hay factores que no dan la posibilidad de tener acceso a los recursos por ejemplo el crimen en la comunidad, desempleo, estatus migratorio, entre otros.

El nivel de crimen en la comunidad es un factor importante al tener acceso a recursos basicos. Por ejemplo, una voluntaria que entreviste respondio a mi pregunta sobre como ella se sentia en la comunidad que ella creció al ir al supermercado. Ella dijo: "Mi familia y yo teniamos meido de salir al supermercado por la violencia que vivia mi comunidad; especialmente por la noche, no era seguro caminar o andar afuera de la casa en general." Por esta cuestion , familias que viven en comunidades donde el crimen es alto no tienen tan facilmente el acceso a comida a comparado a familias que viven en lugares donde el crimen e violencia son bajas. Por ejemplo, yo crecí en

Richmond donde la violencia e crimen se vive a diario. Yo recuerdo que no tenía permiso de salir a la calle por la noche ya que mi madre sabía sobre la violencia e crimen que había por toda la ciudad. Estoy me prevenía asistir a practica de fútbol en mi escuela o tener participación en grupos que se reunieran por las noches. Estos dos ejemplos dan a luz como geograficamente si importa donde viven las comunidades subatendidas e como la geografia es un factor importante que excluye a comunidades subatendidas de recursos basicos como comida e educación. Otra entrevistada respondio: *"De peque*ñ*a me acuerdo que habia ocasiónes donde mi familia no salía por noche por miedo de la violencia en la comunidad. Mi padre llegava de trabajar antes de oscurecer por razones mismas. Yo no podía asistir a la libreria tan tarde; y mi madre no salía a la tienda después de oscurecer.*

El media de transportación es otro factor que geograficamente afecta a comunidades subatendidas a tener acceso a recursos basicos. Yo me acuerdo cuando no tenía media de transportación a practica de fútbol por cuestion de donde viva. *De peque*ñ*o, mi madre y yo, caminamos de mi casa hasta el centro de salud que nos quedaba más o menos hora y media. Un medio de transportación nos habría ayudado mucho porque yo tenía que faltar a la escuela por razon que el entro de salud estaba muy lejos.* Una persona que entrviste me dijo su respuesta a como el transporte fue un factor para ella y su familia para tener acceso a recursos como el centro de salud e educación. Ella respondió: *"Había muchas veces cuando tenía que caminar a la escuela o biblioteca porque no había un metro que nos podría llevar a lugares comúnes. Las carreteras en el centro valle son largas, donde hay campos, lo cual no hay medio de transportación de mi casa al la ciudad."* Ella añadio:*"recuerdo que mi padre tenía que pedir 'raite' a mi tío o compadres para ir a trabajar a la cuidad." El transporte fisicamente puede prevenir a familias, especialmente esas que son subatendias, a poder tener acceso a recursos basicos.*

*Mundialmente, esto también tiene aplicación. Por ejemplo, yo participe en un programa de salud que prov*é *medicamentos a familias que visitan el centro de salud. Para muchos en Honduras, las familias tenian que caminar bastante para poder llegar hasta el centro de salud y poder tener acceso a el medicamento. La mayoria de familias* *tienen que dejar obligaciones para poder a ir al centro de salud. Otras barreras que no dan facil tener acceso a recursos son las condiciones geograficamente donde vive la gente. Por ejemplo, al tener que subir e caminar alrededor de montañas por kilometros, mucha gente sufre de diabetes e asma y dificulta el paso para llegar a el centro de salud. La transportacion es un medio que muchos subestiman por lo cual no hay mucha atencion a esta barrera. Durante el programa, los estudiantes voluntarios como yo ayudábamos en diferentes talleres. Yo tuve la oportunidad de ver a los pacientes que llegan para decir sus sintomas. Alli en el triage yo hablaba en español con los pacientes y ellos me contaban sobre el largo vieja que tuvieron que hacer para poder llegar ese dia al centro de salud. Yo observe que mucha gente llegaba con problemas cardiovascular e fisicamente que daban a luz la dedicacion de la gente a ir tan lejos por recursos medicos. La seguridad en el pais era muy dedicado a extranjeros pero a los ciudadnos no se miraba mucha atencion. Los estudiantes era dados atencion por dos personas del ejercito, y nada de dedicacion para protejer a los ciudadanos. Los pacientes me contaban sobre las insiguridades de la zona y como ellos se sentian muy inseguros.*

El estatus migratorio es otro factor important que muchas veces no es tomado en cuenta. Cuando una familia, por decir, viene hasta los Estados Uunidos, no cuenta con mucha capacidad para enfrentar los riesgos que hay con pedir ayuda a servicios comunes. Yo recuerdo cuando mi familia y yo llegamos aqui en Richmond, CA, y el miedo de buscar ayuda en servicios por la ciudad como para obtener comida a pajo precio y preguntar para poder recibir vacunas por poder entrar a la escuela. Tambien me acuerdo que para obtener servicio en el centro de salud teniamos que esperar una eternidad. Hay casos donde tambien trabajadores se sienten sin poder de demander a sus jefes por miedo de ser deportados y separados de sus familias. Tambien hay veces que las personas indocumentadas son aprovechadas por gente que amenaza por mandarles "la migra". Por estas razones, la comunidad indocumentada no tienen acceso a recursos basicos por razon en donde viven. El miedo runda por las calles y gente sin licencias para conducir prefiere caminar, pero tambien esta comunidad muchas veces viven en comunidades impactadas por el crimen e violencia.

Estoy afecta mucho mas a esta comunidad en tener acceso a estos recursos y ser aprovechados.

Alicia Cuevas: Increasing STEM Education

According to the US Department of Labor, minorities make up less than five percent of the STEM workforce (Williams). Minorities, specifically in underserved communities, are being left behind in the field of STEM education. As Dr. Freada Kapor Klein, co-founder of Oakland-based Level Playing Field Institute puts it, "The nation is suffering from a full participation in STEM" (Williams). It is important to note that the United States is dropping in global ratings in terms of largest global technological economies and proficiency levels in STEM. The nation *must* increase STEM educational resources, especially to underserved communities, in order give equal opportunity in the field and for the advancement of the nation.

We must attack these issues with a sense of urgency. Fortunately, there are programs that have been developed to address these issues and bring STEM education to underserved communities across the nation. For example, the Level Playing Field Institute in Oakland is an organization committed to eliminating barriers faced by underrepresented people of color in STEM. One of the programs, SMASH Academy, takes low-income high school students of color and immerses them in STEM classes on college campuses like UC Berkeley, while also giving them academic support from faculty to ensure continued academic success in higher education. The ACPHS Academy in Albany is an after-school program for third-eighth graders where children are led through demonstrations exposing children to real-world STEM applications. It is refreshing to hear that programs like these are developing, and showing positive impacts.

Although I did not participate in wonderful programs, I had the opportunity to interview a couple of fellow MCB 15 classmates who took part in two different STEM programs in the Bay Area. The first student participated in the SMASH Academy at Stanford University. As an undocumented student, he heard about SMASH Academy through a counselor from another program he was in called College Is Real. According to him, the highlights of the program were that it brought a group of very diverse students directly to the university's campus where they received focused attention and resources to help prepare for college. The most impactful aspect he took away from the program was that it was here that he realized that higher education was in his reach, and he believed that it strongly impacted others around him as well. One aspect that he did not like was the competitiveness in that students needed a 3.7 grade point average to even be eligible for the program. This automatically eliminates those students who may be very passionate about STEM and do have potential, but just do not have the numbers. I feel that programs like SMASH Academy should not look at numbers like GPA, but instead at the actual person to discover if they have the passion and inner potential to pursue STEM.

The second peer that I interviewed was a part of the Upward Bound Math/Science Program at UC Berkeley, a branch of UC Berkeley's Pre-College TRIO Programs. Upward Bound functions much like SMASH Academy in that it integrates underrepresented students into a college campus and provides them with resources to aid them in accessing higher education. However, unlike SMASH Academy, there are no grade point average requirements. The student who I interviewed was inspired to pursue STEM in college thanks to this program, and she also mentioned that many peers continued on to pursue STEM majors as well. In fact, she decided to attend UC Berkeley because of her wonderful experiences in Upward Bound. This program had a great impact on her life, and she is planning to give back to the program by being a residential advisor for Upward Bound this summer. Lastly, she informed me that Upward Bound opened bridges to those around her as her younger siblings are now taking part in the program.

It is important to look at the impacts on students from these sort of programs. Based on the interviews with my peers, I can really see that programs like SMASH Academy and the Upward Bound Math/Science Program are working. They are taking students from disadvantaged backgrounds and giving them the resources and inspiration to pursue STEM. It was amazing to hear how impactful these programs really are for students.

Programs like these *must* continue to grow across the nation. Hands-on exposure, introduction to role models and resources, and partnerships with universities and industries are the best methods to increase the number of under-represented students in STEM. Giving these students these opportunities will lead to them making positive impacts back in their own communities and across the nation.

Work Cited

Williams, Joseph P. "Bringing STEM Education to Underserved Communities." US News. U.S.News & World Report, 29 May 2014. Web. 25 Apr. 2015.

Interviewee One. Personal Interview. 06 May 2015.

Interviewee Two. Personal Interview. 06 May 2015.

Thu-An Tran: Language Barriers and Health Care

Growing up in Oakland, California and coming from a low-income minority family, I experience first-hand what it is like to live in poverty and lack health care. I still remember the many times my parents refused to go to the doctors regardless of how sick they felt just because it was too expensive and "complicated" to go to the doctors. I know that it was not just my parents who experienced barriers, especially language barriers when it comes to accessing good quality health care, but also many other people that come underserved communities with similar backgrounds. In addition, it is not just the patients, but also medical professionals that are affected by language barriers. A study published in 2006 "demonstrates that acute care hospital medical professionals perceive language barriers as an impediment to quality care delivery and as a source of workplace stress" (Bernard et al., 2006). In this paper, I want to discuss the language barriers that exist in underserved communities when it comes to assessing good quality health care.

According to the 2011 U.S. Census, of 291.5 million people aged 5 and over, 60.6 million people (21 percent of this population) spoke a language other than English at home (Ryan, 2013). Overall, 15 percent self-reported that they spoke English "not well," and 7 percent self-reported that they spoke English "not at all" (Ryan, 2013). Also, speakers of all languages other than English who spoke English less than "very well" had not changed as a percentage of the total population 5 years and over from 2007 to 2011 (8.7 percent) (Ryan, 2013). This percentage had increased from 8.1 percent in 2000 to 8.7 percent in 2007 (Ryan, 2013). This diversity in languages that exists in the U.S. makes communication between individuals more difficult. If an individual does not speak English, he/she will not be able to or will have a very hard time when it comes to accessing the available resources that are provided to people who live in this country. This is especially the case when it comes to accessing good quality health care, at least from my own experience.

I came to the United States from Vietnam with my father and my older brother when I was 11 years old; my mom had decided at the time to stay in Vietnam with my little sister. Three months after I have stepped foot in the city of Oakland, California,

I immediately started middle school. Since I was at the age of puberty, my menstrual cycle started during my first year in middle school. As the only girl in the household at the time, I felt too embarrassed to tell my father and brother about the situation. To make matters worse, I could not talk to my mom because she was in Vietnam, so I did not know what was going on or what to do. Fortunately, there were resources at my middle school, more specifically a clinic to help students there with any medical concerns. Unfortunately, I did not know any English; thus, I was not able to communicate with the staff there to ask for help. It went on for days until one of my older Vietnamese friends finally noticed and explained to me what was going on and what I needed to do. It sounds funny now looking back at it, but at a young girl at the time, I honestly did not know what was going on with my body. I felt alone because I was not able to communicate with anybody, even though I know that the resources were there. The whole experience really distracted me from my schoolwork and my social life, in a negative way.

In an underserved community like Oakland, there are many immigrants coming from many places around the world. Immigrants or individuals with an immigrant background, like myself face many challenges when it comes to accessing good quality health care, due to language barriers. I know there are services being offered by the health care system to individuals who are non-English speakers, such as interpreter services or having available resources (i.e. pamphlets, brochures, etc.) in different languages. However, even with these available services and resources, problems can still arise. For example, while there are available resources, such as pamphlets or brochures in different language, most of the resources are only translated in common languages and not in less common languages. We need to increase the number of languages being used for translated-resources in order to reach all non-English speakers. Rita Webb, a social worker stated, "Language barriers play a significant role in how health and mental health services are delivered, received and understood" (Webb, 2010). Here, it is important to note that not only do we need to provide resources for patients (people from underserved communities), but it is just as important for health care providers in these underserved communities to have resources to help them deliver better health care to the people they are serving. Personally, I think that the best way to do this is to train more health care professionals who come from underserved communities, who understand the challenges that people from underserved communities have to face due to language barriers, who speak languages other than English, and most importantly, who want to support and help provide good quality health care to patients that face language barriers from underserved communities. Having health care professionals that really care about the well-being of underserved people will allow for more effective and efficient health care.

Despite all I have said, we need to look at the root cause of this problem regarding language barriers. At the end of the day, it is the inability to speak English that creates the language barriers that exist in the U.S., which leads to miscommunication and lack of access to available resources. Ergo, it is important to recognize the root cause and come up with a solution to address this problem of language barriers. Personally, I believe that providing education to everybody in underserved communities, such as English classes is a solution to this problem of language barriers.

Just like many other immigrant families, my family comes to the U.S. to seek for opportunities. My parents left everything and came to the U.S. in order for their children to have a better future and education, despite the fact that they would have to start everything over from scratch. I was blessed enough to be able to attend UC Berkeley, the number one public University in the world! I have been learning so much during my past three years here. Being here at UC Berkeley, I have recognized the importance of education and how critical it is to give back, especially to my underserved community at home. My future goal is to become a health care professional to provide the best health care to people in underserved communities, while helping address the problem of language barriers. I know that if anyone that faces language barriers is given the opportunities that I have been given thus far, he/she will also recognize the importance of education and how critical it is to do everything in their power to improve the lives of people from their own communities, through different means. Link to video: https://youtu.be/RpapxCxs_kA

References:

Bernard, A., Whitaker, M., Ray, M., Rockich, A., Barton-Baxter, M., Barnes, S. L., ... Kearney, P. (2006). Impact of language barrier on acute care medical professionals is dependent upon role. *Journal of Professional Nursing: Official Journal of the American Association of Colleges of Nursing, 22*(6), 355–358. http://doi.org/10.1016/j.profnurs.2006.09.001

Ryan, C. (2013). *Language Use in the United States: 2011* (American Community Survey Reports No. ACS-22). U.S. Census Bureau. Retrieved from https://www.census.gov/prod/2013pubs/acs-22.pdf

Webb, R. (2010). *Social Workers: A Bridge to Language Access Services* (p. 4). National Association of Social Workers. Retrieved from http://www.naswdc.org/assets/secured/documents/practice/clinical/WKF-MISC-47310.DiversityPU.pdf

Anonymous: The Undocumented

A runny nose, sprained ankle, or aching back are common reasons indicative of the need to visit a doctor's office. It is almost second nature to think of a physician when one encounters these rather trivial, minor health issues. However, this privilege of having plenty of medical resources is not one that has been bestowed upon us equally. On the contrary, there are sectors within one's own local community that experience substantial inequality and lack the ability to seek care. This issue resonates personally since a large portion of my childhood and adolescent years were spent as a member in a community where many of my peers, family members, and myself were members of one these underprivileged groups – the undocumented. I first handedly witnessed a range of scenarios, from the helpless frustration brought on by the lack of resources, to an underlying fear of the legal repercussions that could stem from a simple hospital visit.

As members of this quickly growing population we found ourselves with a lack of resources. Government programs such as Medicare and Medicaid, whose goal is to spread the availability of healthcare to marginalized populations, ironically did not apply to us simply because we lacked a document validating our residence in this country. Why the prospect of a simple doctor's visit was out of the question was a difficult problem to wrap my head around as a child.

Frustratingly, the absence of resources couples itself with its equally debilitating partner – fear. The undocumented of this country have everything to lose at any moment's notice. For example, an otherwise simple traffic violation has the potential to become the medium through which a family is torn apart. With stakes as high as they are, the thought of walking into an institution such as a hospital is a chance people will not take. The few resources that are available are then hurt by the fear and angst present. This all causes people to be less likely to go out and look for few resources available to them.

Hence, that is how the perpetuated symbiosis is constructed – one that is between the lack of resources and the distress that comes along with the sheer thought of looking for help. Undocumented immigrants in this country have been put in an incredibly unfavorable situation, one that takes away their ability to seek healthcare. All ten million of them are at the mercy of either an immigration or inclusive healthcare reform. In the meantime, the cycle will continue to be prolonged.

The current eleven million undocumented immigrants are without resources to combat their marginalized state. The newly indoctrinated Affordable Care Act excludes the undocumented population and they are unable to access health insurance coverage. Ironically, this piece of legislation was put in place to extend the health care coverage to more people in the United States; however, the undocumented was yet again marginalized through their exclusion. As stated previously, not only does this new program implemented by the Obama Administration not include those without proper documentation, current pieces of legislation like Medicaid and Medical are also exclusive of the undocumented. According to the Washington Post, "more than 100,000 immigrants who bought health-care plans through the federal insurance exchange will have their coverage cut off at the end of the month" since they were not able to provide documentation of their residency of citizenship status. It is clear that there are many obstacles and hurdles set in place in order to obstruct undocumented immigrants from obtaining healthcare coverage in the United States. Interestingly enough, the health of the average undocumented individual is better than the average American when first coming to the united states – however, as their time in the states increases, their health deteriorates. Even with multiple studies having proved this – there are still only a few mediums through which these individuals can access some sort of relief.

When exploring the limited resources that are available to those who are undocumented, it is important to begin the discourse with a discourse on how fear is propagated. As sated before, the undocumented of this country live in a state where their place in this country could be removed by a simple traffic stop. A sense of fear is spread in the undocumented communities through government mechanisms. Therefore, the few resources that are available are not always pursued because of this ongoing vigilance that they must maintain.

Many of these individuals are not aware of the fact that they are available to be seen in an emergency setting – since they are afraid of government retaliation. However, through the Emergency Medical Treatment & Labor Act, emergency rooms are required to provide care to anyone regardless of their residency status. It must be noted that even when the undocumented visit the ER – this is as a last resort when things have gotten to a critical condition. This only feeds our already existing, broken healthcare system where we only treat diseases and we do not implement enough preventative care.

The silver lining comes in the shape of community health centers and other sort of nonprofits. However, as previously stated – the spread of this information is damaged by the fear of the consequences that could come from seeking help. These options are there for them to receive primary care and prevent chronic illnesses from becoming a problem. The centers are – Federally Qualified Health Centers that receive their funding by the federal Health Resources and Services Administration. I have the pleasure of volunteering at one of these clinics that serves the uninsured and undocumented populations of San Francisco. Through my service in such a charitable service provider – I have seen the importance of serving those who are marginalized by government regulations that exiles them because of their residency status. The need to care for people who lack resources is apparent in counties and cities across our country. Unfortunately, as we continue to expand healthcare resources to the average "American" we continue to further the oppression and marginalization of an already underserved community – the undocumented.

Works Cited

"Emergency Medical Treatment & Labor Act (EMTALA)." CMS.gov. Center for Medicare and Medicaid Services, n.d. Web. 10 May 2015.

"Emergency Medical Treatment and Active Labor Act." Wikipedia. Wikimedia Foundation, n.d. Web. 9 May 2015.

Gusmano, Micheal K. "Undocumented Immigrants in the United States: U.S. Health Policy and Access to Care." Undocumented Patients - Undocumented Immigrants and Access to Healthcare. The Hastings Center, 03 Oct. 2014. Web. 10 May 2015.

"Immigrants And Health Care: Sources Of Vulnerability." Health Affairs. N.p., n.d. Web. 10 May 2015.

Sun, Lena H. "115,000 Immigrants to Lose Health Coverage by Sept. 30 Because of Lack of Status Data." Washington Post. The Washington Post, 14 Sept. 2014. Web. 9 May 2015. "The Uninsured: Access to Medical Care Fact Sheet." American College of Emergency

Physicians. N.p., n.d. Web. 9 May 2015.

"Who Is an Undocumented Immigrant?" Nolo Law for All. N.p., n.d. Web. 11 May 2015.

Dawa Gangshar: Improving Nutrition, Reducing Obesity

Here in the United States, obesity has been significantly increasing and disorders associated with it have also been on the rise (Sturm and Hattori, 2013). According to the Centers for Disease Control and Prevention, African Americans are 51% more likely to be obese than Whites, and Latinos are 21% more likely. Understanding circumstances that people in underserved communities face is a step toward promoting healthy eating and reducing obesity rates. Circumstances that said populations face are largely based on the lack of crucial resources. Resources we may take for granted, such as a local supermarket or even accessing food vouchers to purchase healthier food options.

Many underserved populations consist of families living at or below the poverty line and may find that purchasing healthy foods are expensive – though some studies show otherwise. In a 2008 public health study, low-income women were given redeemable food vouchers for fruits and vegetables. The results proved successful, as many of the participants increased their consumption of fruits and vegetables far higher from their baseline diet (Herman et al., 2008). This study highlights the poor food choices made by those who are low-income in underserved communities. However, with proper and necessary interventions, a family's diet can be remarkably reversed and improved. Another example of such intervention is the installation of a local supermarket in underserved communities. A recent study showed supermarket and grocery store interventions help in influencing healthier food choices, however, there must be a demand for these interventions on the part of the consumers (Escargon et al., 2014).

I believe proper nutrition education in addition to access to resources plays a significant role in food choices made by members of an underserved community. Though many circumstances can affect one's ability to receive a proper education, such as the lack of transportation or mentorship. I also believe in the significance of instilling community leaders who promote good nutrition and leading a healthy lifestyle. These leaders are passionate members who are part of the community themselves and have the adequate knowledge in nutrition and teaching skills to bring about improved changes.

Without resources and the knowledge on maintaining good health, underserved populations are at risk for developing unhealthy eating habits. This can increase the prevalence of obesity-related health issues, which can put an even greater burden on these communities. Thus, proper resources and implementing effective interventions, such as access to local supermarkets or receiving food vouchers to purchase fruits and vegetables, can significantly influence healthy food choices.

References:

CDC (2009). Differences in Prevalence of Obesity Among Black, White, and Hispanic Adults --- United States, 2006—2008. *CDC MMWR.* July 17, 2009/58(27);740-744. Retrieved from: http://www.cdc.gov/mmwr/preview/mmwrhtml/mm5827a2.htm

Escaron AL, Meinen AM, Nitzke SA, Martinez-Donate AP. Supermarket and Grocery Store–Based Interventions to Promote Healthful Food Choices and Eating Practices: A Systematic Review. [Erratum appears in Prev Chronic Dis 2014;11. http://www.cdc.gov/pcd/issues/2014/12_0156e.htm.]

Herman DR, Harrison GG, Afifi AA, Jenks E. (2008). Effect of a targeted subsidy on intake of fruits and vegetables among low-income women in the Special Supplemental Nutrition Program for Women, Infants, and Children. *Am J Public Health.* 2008 Jan;98(1):98-105.

Sturm R, & Hattori A. (2013). Morbid Obesity Rates Continue to Rise Rapidly in the US. Int J Obes (Lond). 2013 Jun; 37(6): 889–891.

Part 3: Language and Cultural Barriers

The goal of our part of this chapter is to analyze the cultural and language backgrounds, beliefs, and morals to understand barriers and their origins.

Glendy Ramon: Language and Cultural Barriers within Latino immigrants

To understand how language and cultural barriers become exposed to many communities, I analyzed the many factors in which prevents low income Hispanic immigrants from my community to continue in comprehending scientific topics both in articles and news, especially concerning about medicine and health issues in today's news.

After talking to several people from my neighborhood, mostly mothers, some explained that they were afraid and ashamed to ask doctors about certain health issues. They were uncomfortable to ask doctors and nurses more information about certain health issues, therefore preventing them from asking questions and just agreeing and following whatever they said. Some due to not talking English as well as a lack of language translators in clinics were unable to communicate fully making it hard to understand from both sides, which caused a loss of emphasis on scientific terms due to translation. One event someone had to go through was that the doctor "criticized and humiliated [her] for not understanding or having enough knowledge about the health issue which she must be responsible in knowing". Being so, most people in my community have a hard time educating themselves if it's either through technology or going to the library. Most people in my neighborhood do not own a computer and those who do have a hard time using it. Most rely on having their children to research things and letting them know the information. I had to do it several times for both my parents and it was a challenge as well when I tried explaining things for them in which I didn't even understand when I was younger. It was easy for the significant terms to lose meaning and focus as I tried translating the topics to simpler terms.

The public library also has some difficulty in providing resources for the Spanish-speaking people from the community. There is a small section in the library that has books in Spanish but even so, most of these books are mostly stories, magazines, or soap operas. There is rarely any scientific book or article that allows them to learn more about medicine and science overall. Many are also old books, which prevents readers to know of the progress and development in science throughout the years therefore bringing another barrier to most of the Spanish speaking community.

All the people who I talked to are immigrants from Latino countries who had to adjust or understand US culture. However, many kept traditional views from home preventing them to accept certain science topics and medicine affects. Whenever my friends from the neighborhood and I got sick, instead of being taken to the doctors many mother's and neighbors made us go through house remedies from their native homes, if it was either a special food recipe or spiritual healing. It shows that even with doctor's suggestions and instructions to a certain sickness, there are still some who refuse to listen and continue their own remedies and treatment due to it being passed from generations before them. Especially in my neighborhood, where most of the Latino immigrants come from some villages and cities with little to no clinics or hospitals they rely on what they had been taught from parents or those who had a bit of experience curing others.

Joanna Ruelas: The Need for Diverse Health Care Providers in Latino Communities

Communicating science to low income community is always something very difficult to do. This becomes even harder when you bring in differences in language and cultural beliefs. My goal was to analyze Latino communities with a focus in the health care setting and try to understand how certain language barrier within science came to be and the impact they have on the overall public understanding of science. Over the course of my life I was needed as a translator for my parents which exposed me to the barriers within the healthcare setting. These healthcare interactions are very important because they shape the perspective that many people have of science, since in many cases the general public gets many of its first exposures to science through the health care setting. The lack of proper communication and different cultural belief has led to movements such as the anti-vaccination movement which can have potential negative impacts on the community as a whole. Issues like these could be prevented if there was more cultural humility between those in the scientific field and those outside of it.

I interviewed some individuals from my Latino community to see what their experiences have been in the healthcare setting. One individual explained her experiences going into the clinic for her constant pregnancy checkups. She stated that she did not know any English and was too scared to ask any questions because she could not properly express herself, causing her to leave confused about what the doctor had just said. She stated when staff tried to explain information to her, she felt they were just reciting out of a book and did not find it helpful at all. She stated that she found more help talking to women who had gone through the same experiences she was going through. Interactions like these show the failures of the healthcare system because they neglected to address the patients' needs simply because the doctor could not communicate with the patient.

While many clinics have interpreters to help patients understand what the doctor is saying, by having that third person in the room it breaks the bond between the doctor and patient. This is not the only issue however, in many cases interpreters are present but only confuse the patient even more. I have been a witness at times when an interpreters were assigned to my mother. In one instance, I remember a doctor going into detail, trying to explain what was going on; however, when it came time for the interpreter to explain what the doctor had said, he simply stated, the doctor said your fine. He had neglected to state the explanation that the doctor had given, leaving my mom confused as to what was going on with her health. Due to these experiences, my mother would prefer to have me interpret for her, even if I was just a child, because she felt more comfortable asking me to ask questions and trusted me to explain what the doctor was saying. Through all of this brings up the notion that just having interpreter services does not decrease the gap between doctor and patient but rather it is the quality of the interpreter services that really help make a difference

Experiences like these show that there needs to be a change in the way that the healthcare setting is run. To help fix the gap between doctor and patient communication we must diversify the doctors going into the healthcare field so that patients find a better connection with their doctors as well as improve cultural competency in the healthcare setting. Placing doctors in areas where they connect with their patients will have a stronger impact on the overall health of the community. There have been experiences where doctors are placed in an area where they do not wish to work reducing the quality of care given because they do not understand the background of the community. Especially in low income communities it is important to choose the right staff. One interviewer noted that a doctor once attended her but took less than five minutes with her and began to walk away even before she was done talking. She stated that he gave off the impression that he felt superior to the patients he was attending. In order to avoid situations like this there needs to be more cultural humility in the healthcare setting to better fit the needs of the community. However, this does not only apply to the medical staff, this as well must be addressed with the interpreters, receptionist and anyone else working in the health field. Staff must as well be diverse and properly trained in their jobs in order to improve the quality of care given to the patient.

Maricarmen Hernandez: Language and Cultural Barriers: Mediating Barriers in Health Care Settings

Going to the doctor to seek care is common experience for everyone. However, to patients who do not understand the language their health care provider speaks can be a stressful and uncomfortable experience. As an immigrant family, my parents and I have first-hand experience of the difficulties one can experience when communication with a health care provider is diminished due to language barriers. When I was thirteen years old, I was diagnosed with a jaw tumor and required multiple medical procedures and surgeries. During that time, seeing the doctor, or even going to the hospital in general was a stressful experience. My parents did not speak English, and at times due to my condition, I was not able to speak. The only way to advocate for myself at times was to write things down, and that slowed things down in regards to receiving care. The times when I could speak, the difficulties were properly advocating for myself, since I was very young and did not understand how the health care system worked. This frustration then had a rippling effect and affected the relationship I had with my parents.

After months of treatment and going in and out of the hospital, there was one visit that I still vividly remember after 10 years. I was at my pre-operation appointment and the nurse was going to assess my health to be cleared for surgery. She noticed my mother did not understand English and as I was about to tell her that I would interpret, she told me I would not have to worry and an interpreter would be called by phone. Such a simple and logical statement brought so much relief to me in that period of my life that I still remember it after a decade. I am certain that many other patients with limited English proficiency or who have family with limited English proficiency have been in my shoes and did not know an interpreter could be accessed during a hospital visit.

Numerous studies have documented the profound adverse impact of language barriers in regards to access and quality of care[1]. Patients with limited English proficiency who need but do not receive an interpreter have the lowest satisfaction of care compared to any other group of patients, and even when using ad hoc interpreters, such as family members, they are still much less likely to be satisfied with their medical visit[1]. Additionally, many studies document the positive impact of trained professional interpreters. Patients with limited English proficiency who are provided with interpreters make more outpatient visits, have better health outcomes, and have higher satisfaction with care[1].

Unfortunately, trained interpreters are often not offered in all health settings and some insurance plans do not cover cost of interpreter services. Additionally, in some places that do offer interpreter services, they are often not called into action due to time constraints experienced by physicians. Obtaining an interpreter can take time

---

[1] Ku, L., & Flores, G. (2005). Pay now or pay later: providing interpreter services in health care. *Health Affairs*, 24(2), 435-444.

out of the already short time physicians have with their patients, and interpreting itself takes additional time as well.

Some methods to improve interpreter access and quality of care include using registration systems that link directly to interpreter scheduling systems appears to decrease wait times for interpreters[2]. Furthermore, formal training in medical interpreting along with assessment for language fluency can help mediate instances in which medical interpreters can make mistakes, which may have serious consequences for patients. Lastly, we know the great impact interpreters have on patient care and access; however, this service is only useful if it is demanded. Securing the services does not promise that patients who need them will get them. Like in my story, after months of frustrating doctor visits, we were finally told during a visit with a nurse that interpreter services were available. Therefore, without demand from physicians and nurses, the supply of interpreter services can go untapped or underutilized. Thus, the easier it is for these providers to obtain the language service, the more likely they will include it as routine practice.

Therefore, when you or a family member require a medical interpreter when receiving care, please make sure to request one. If medical interpreter services are offered at your health care provider's, please do not settle for ad hoc interpreters. It is within one's right to request an interpreter and to be provided a qualified interpreter where this service is provided.

Anthony Meza: How Economic and Language Barriers Prevent Underserved Communities from Accessing and Engaging in Science

In the section "Language and Cultural Barriers", the overall goal is to analyze how culture/language backgrounds, beliefs and morals can lead to barriers that impede underserved communities from understanding science and how this problem originates. In terms of my write up, I am looking at the specific language/ economic factors that impede my Spanish-speaking community from engaging in scientific

---

[2] Regenstein, M. (2007). Measuring and improving the quality of hospital language services: insights from the Speaking Together collaborative. *Journal of general internal medicine*, *22*(2), 356-359.

activities, as well as, accessing scientific journals to be more knowledgeable about modern day science. In order to properly assess where my community stands, I took a mini-survey of about five people and asked them specific questions such as "What does science mean to you", "Are you interested in learning science". I would then follow up with questions such as "Why are you (or not) interested in learning about science" in order to reach a conclusion of what does "understanding" science mean to my underserved community.

Upon conducting the survey, three of five people acknowledged science was extremely significant for the good of the population whereas the other two people saw science as a "bunch of ideas that only the educated seem to understand". When I asked my interviewees if they're interested in learning in science, an astonishing four out of the five said they would not be interested. The one person that was interested in learning about science emphasized that "knowing science is the way to live and because I do not have kids, I can focus my energy and attention into learning something that will benefit me".

However, it was very interesting to hear the reasoning of the other four folks of why they would not be interested in learning about science. A big reason of why science was not appealing to them was due to their hectic schedules during the day. These folks along with their family members have two and even three jobs, in which they work six days a week. This itself consumes their time and energy, which is bad if one wishes to be engaged in science; this is because in order to be engaged in science, one needs time, dedication and energy. However, an interesting note I took from one of the interviewees was that if it weren't for the hectic schedule of washing cars by day and dishes by night, he would love to learn about the scientific terms that are being said on the news. Another big factor that impedes my Spanish-speaking community from accessing or understanding science is the language barrier. They feel that because they don't know English, they are deemed to be unworthy of learning about science and as a result, they don't pay attention to it.

As of now, there exists a few solutions that help out my fellow community members be more engaged in science. For example, in terms of the language barrier, the public library offers Spanish books and articles relating to science, in which anyone could stop by and read if they wish. In terms of the economic barrier, there exist some late night group meetings where people can show up after work and talk about current events (relating to science). Although these programs appear better on paper than actually helping my community, it is good to note that these programs exist for a noble cause.

However, a successful solution that might help my community members overcome the economic/language barriers in order to "bridge the gap" of their understanding of science is to focus on the children and motivate them to learn about science. Why the children? Parents live with their children and constantly interact with them, whether it is through meals or in car rides. If children are excited about science, then they would most likely talk to their parents (in the household's native language) about what they learn about science in school, and thus educate their parents and even motivate them to use their leisure time to pick up a book and learn on their own.

Claudia Pacheco: Latino Faith and Medicine

When people refer to the "public," who exactly defines this group? There is no way to uniformly characterize this group and often we have to piece apart the population that makes up this group when analyzing the public understanding of science. One piece of that public is the Latino community, and to no surprise this group is also widely varied. One general way to characterize this piece of the puzzle that is the public is through a shared culture and often set of beliefs. To focus the discussion even more I want to touch upon how these ingrained beliefs and values are influenced by one particular aspect of culture, namely religion, and how this affects specifically how health care and science is observed.

Religion develops culture: Religion has influenced the development of Latino culture and traditions. Often many of the common phrases heard are equated with religion like "si Dios quiere[3]" to express hope in an outcome, "bendito sea Dios[4]" when something has gone well or "Dios mio[5]" when surprised or feeling of anguish. Many of the traditional Latino celebrations and holidays are also religious like Holy Week (Cuaresma) or Three Kings Day (Los Tres Reyes Magos). In the same way, the virtues and morals of religion have served as main players in shaping perception about family, education, values and therefore science. It is hard to break down Latino culture away from its religious influences and similarly it is hard to find the intersection between religion and medicine but it is a critical area to be explored so that the gap between a patient and understanding their treatment and health.

How culture influences science perception: Focusing specifically on the health science aspect of the discussion and keeping in mind the types of beliefs shaped by the Latino culture, the most prominent way in which science is perceived is often through a lens of faith. Often healing and health is put in the same category of spirituality. Not specifically leaving all the outcomes and action up to God or a higher being, but putting faith that whatever happens is supposed to happen or if it did happen it is a direct reaction to one's own actions or character. The dictionary defines science as "systematic knowledge of the physical or material world gained through observation and experimentation." The explanations for the observation of the world are already answered through faith and culture. Therefore, when new ideas or discoveries come about, especially without proper communication and explanation, results in individuals not accepting or brushing off an idea as crazy or wrong or in our case medical interventions.

Outcomes of not addressing the barriers and how to address them: By not considering culture when communicating science in the healthcare setting, it results in many people putting their faith in folk healing or word of mouth treatment. By not understanding, they lose their faith in doctors and medicine and like many of my family members they stop taking medications or following treatment. This can create a challenge in coordinating health treatment with a physician and can lead to

---

[3] If God wants
[4] praise the lord
[5] My God

worsening health. Modern medicine is many times given the label of "evil" and untrustworthy, which is a belief that could be alleviated if health promotion and health science was more culturally appropriated. Science needs to be spoken in terms of how it influences behaviors, attitudes and knowledge, which is the basis of culture.

After having conversations with a few devout Latina Catholics and also friends, one particular story stood out to me. Let's call this friend, Rosa and her mother Refugio. Along with this narrative, let us think about a familiar topic in the religion vs medicine debate: Contraception.

Rosa had just gone to the doctor and due to some hormonal imbalances her physician decided to prescribe birth control. Rosa calls her mom, Refugio, and tells her the treatment because let's face it, Latina mothers always need to know what is happening in their children's life. Refugio, a Catholic Latina, automatically gets angry and says definitively that she won't be taking the medication and that the doctor is just crazy [insert Latina mom scolding rant here]. Rosa, an educated Catholic Latina listens to her mother's angry rant and understands where the mother is coming from in terms of their religious beliefs, but she also understands why the doctor has prescribed this medication because of the biology courses she has taken. Refugio assumes that taking birth control means the doctor must somehow be promoting sexual activity before marriage or just recognized the negative connotations in Catholic culture when it comes to birth control.

From this very brief narrative, one can pick out a few critical observations. One was that Rosa had no objections to treatment because she recognized what the medication did beyond what it has been stigmatized as and that led to a prompt start of treatment. What Refugio didn't have was prior knowledge and an explanation about the function of birth control and therefore treatment of Rosa's imbalance. Refugio instead analyzed it through a lens which she was knowledgeable about, and therefore could not comprehend this treatment because she only knew the explanation from the side of Catholicism that strictly prohibits its use.

This is a similar story for many Latinos, often they do not know why a certain treatment is given or they do not understand a basis for it and therefore instills mistrust in the health care system. Modern culture accepts many things as a given, such as birth control, and therefore professionals many times don't find themselves in need for their patients to understand their decisions as well.

How can the intersection of Latino culture and medicine be mediated: A critical area that health professionals need to address is being aware that cultural beliefs can shape how an idea or piece of information is grasped. A focus should be placed on training physicians and medical staff to be culturally competent. Being educated and having the power of knowledge means that they or you are placed in a position to educate others. To do that one needs to understand how to effectively transmit information. Only then can we bridge the divide between cultural belief or faith with medicine.

Ruvim Kushnir: Mistrust in the Community

Along with the language barrier, members from the Soviet community are also familiar with the propaganda and censorship in the former Soviet Union, which makes it difficult for them to believe anything they hear from American academia. They constantly mention "it is just the government supplying researchers with money that allows them to make up information just so they can earn more money." Statements like this one are statements that I grew up hearing regularly. By learning about the scientific community, I aim to reach out to people in my community to gain a deeper understanding of why they are reluctant to believe information that the scientific community presents and what could be done to gain their trust. In addition, I also want to speak to some of them and have a conversation with them regarding this matter.

After speaking with several people from my community I realized that they are actually interested in learning about the information that researchers in the scientific community have to say about the efficacy of medicine, nature, vaccinations, or American academia in general. The main problem that arises is that they hold conflicting views of Soviet and American academia that makes it difficult for them to trust the information presented in America. The most interesting conversation I had was with an older man from my church who recently emigrated from the Ukraine. I will provide some of the highlights of the conversation in Russian, translate it into English and then elaborate on it.

(Me) Как ваш опыт в качестве иммигранта был в получении Медицинское лечение здесь?

(Answer) Они не заботятся обо мне! Каждый раз когда я иду к врачу я никогда не смогу объяснить мои осложнений правильно так как он не понимает меня и лишь изредка они когда-либо предоставить мне переводчика. Я говорю им что постоянно болит печень и что я не могу принять определенные лекарства, но они предписывают мне это независимо. Они просто хотят денег.

(Me) Почему бы вам не довериться профессионалам здравоохранения?

(Answer) Вы знаете, во времена Советского Союза, это были настоящие врачи, которые действительно заботилис о каждом пациента. После каждой операции, или даже рождения, пациент был не в состоянии идти домой в течение двух недель, так как они хотели чтобы убедиться что пациент полностью выздоровел. Чтоби они не ицпытали осложнения и вернулись. Но все врачи в Америке только желают заработать как можно больше денег. Они пытаются пнуть пациента сразу после операции. Какой вид лечения является ето?

(Me) Когда эта идея начала возникать, что они просто хотят денег?

(Answer) Вы шутите, весь мир знает об этом. Я вырос в слухе, что американцы будут делать что-либо за деньги и что капитализм плохая экономическая система. Радио и телевизоры постоянно напоминали нам, как жестоко это форма экономической системы.

(Me), вы верите в научном сообществе и литературе они обеспечивают?

(Answer) Я не знаю. Я не обращаю на это внимание, столько сколько раньше в Советском Союзе. Так как я не знаю если они даже предоставить корректную информацию, они могут легко манипулировать информацией чтобы заработать больше денег.

(Me) How has your experience as an immigrant been in obtaining Medical Treatment here?

(Answer) They don't care about me! Every time I go to the Doctor, I can never explain my complications properly since he doesn't understand me and only seldom do they ever provide me with a translator. I tell them continuously that my liver hurts and that I cant take certain medication but they prescribe it to me regardless. They just want the money.

(Me) Why don't you trust the health professionals?

-> You know, during the Soviet Union, those were the real doctors who truly cared for each patient. After each surgery, or even birth, the patient was unable to go home for two weeks since they wanted to make sure the patient recovered completely so they wouldn't experience complications and return. But all the Doctors in America only desire to earn as much money as possible. They try to kick the patient out immediately after the surgery. What type of treatment is that?

(Me) Is it difficult for you to trust American science?

(Answer) How can I trust it if I don't understand it? And who knows if they didn't just make something up.

(Me) Do you believe in the scientific community and the literature they provide?

(Answer) I don't know. I don't pay attention to it as much as I used to in the Soviet Union since I don't know if they even provide correct information, they could easily manipulate information just to earn more money.

Based on some of these common Soviet immigrant responses, it becomes very difficult for anyone growing up in this type of community to fully immerse themselves in academia and have any trust in the information that science communicates. People who come from these communities have to take a stand to question certain cultural norms in order to find what is right based on evidence. Furthermore, by paving my own path in American academia I aim to reach out to the younger generation in this community to advise and lead them towards evidence based knowledge instead of statements of distrust, due to previous cultural and social upbringings. Consequently, this will slowly break those barriers set up by the "Soviet community members" and reveal that science is something that can be trusted, as long as the evidence supports it.

Part 4: The Demand of Non-Scientists to "Understand Science"

The goal of this group is understand the motivations and factors that influence the public to learn/understand science. Particularly, we will look at the role of social media, TV, family, health and psychology have in addressing this issue.

Jennifer Diaz: Medicine as a Key to Motivation

Science exists everywhere in our world today. It is part of our daily lives, from cleaning and driving, to recycling and using a computer. Incredible advances in technology and science are transforming our world at an incredible pace, not only improving the way we live but also making a big difference in our environment. Being "science literate" will no longer be just an advantage but an absolute necessity. The motivations for the public to become "science literate" come from all sorts of directions ranging from social media to criminal profiling and so on. Overall the goal of my group was to better understand different motivations that drive the public to want to understand science.

Currently, **7.6 million people** die from cancer worldwide every year, and the war with cancer continues today. The war on cancer has created a great excitement within the science field and the public, families all over the globe are affected by diseases like cancer and this makes them want to understand diseases and find cures. One person that I interviewed was personally affected by Cancer because her aunt passed away from cancer and because of that today she aspires to study science and learn more about human biology in hopes of finding a cure to cancer and putting an end to all the suffering that people are enduring because of Cancer. Personally I have also witnessed something similar in my own household because my brother was born with Spina Bifida; before my parents had my brother they were not interested in science at all but since my brother was born they have found a great passion in trying to understand what science is to better treat my brother and help him in the best ways they can. The interest my parents have for science would have never existed if it were not for my brother. The other person that I interviewed said that she has never had such a personal experience or story that has brought her close to science and finds no happiness in studying science. Although it is not always the case that a personal experience often strikes the interest for non-scientists to understand science I found that medicine and disease could often become a motivation for the public to want to learn about science in order to make a difference in the lives of some of their loved ones.

Claudia Tse: Harmful Effects of Ignorance and Disinterest

Not everyone shares the same level of interest or knowledge in science. Some people's interest and knowledge may vary depending on the environment they grew up in, the education they received, and the people they were exposed to. The importance placed upon learning and understanding more about science is also influenced by these factors and by one's own perception of science. While a person's interest or lack thereof in science may seem personal, their ignorance and unwillingness to educate themselves with the resources given to them can have a damaging effect on their own personal lives and the lives of people they are connected to.

The recent rise of technology has had a strong effect on the power social media has over the public. It has become easier, faster, and more efficient than ever to share news. There are advertisements posted all over bulletin boards, train stations, and even in many of the technology devices we use daily. New information is shared on a multitude of social media webpages such as twitter and reddit every minute. Additionally, there are TV shows that are dedicated just to educating the public on a variety of topics from current news to physical science. It seems as though the public is constantly bombarded with new information.

However, it's important to note that not all of the information being fed to us is necessarily correct and in fact, some of them may even be harmful. Certain talk shows such as Dr. Phil are notorious for spreading erroneous, misleading health information to the public. Advertisements for medication should be looked at with a critical eye. But it's hard to catch all the incorrect information being thrown at us, especially when you lack the knowledge and ability to accurately predict the truthfulness of the information. It is therefore in the best interest of the public to learn how to differentiate between fact and fiction and to better educate themselves in the areas of knowledge that they are harmfully ignorant in.

However, one's lack of interest in understanding and learning more about the different areas of science creates a barrier. My group interviewed two groups, non-science major and science majors, on their interest in science. For most of the science majors that were interviewed, they expressed an interest in learning and studying science simply because the topic seemed interesting and engaging to them. However, for those that weren't studying science, they seemed only interested in science for the benefits science provided to the world and to them personally such as improvement in the quality of life, health, and technology. Judging from the responses received, I think those non-interested in science would be motivated to learn more if they knew how being educated in specific areas of science could affect their lives. They would be more willing to learn about a subject matter if they knew why it was worth knowing. And for those who are already interested in science, they would be motivated to discover different areas of science if that knowledge was shown to them in an engaging fashion.

Although we can't get rid of the harmful, incorrect knowledge presented to the public, the media can play a part in motivating people to educate themselves about science. Hopefully being knowledgeable about the different areas of science will lead to making better, informed decisions.

Emani Harris: Is Understanding Health Enough?

A hot topic that has always been a part of science is health. It's what gets funded; it's what gets shared through cultures; ultimately, it's what the general public goes out of their way to understand. Non-scientists are interested in health because there is a clear reward to understanding it, longevity. Other genres of science have less clear rewards, and thus, the public invests less time to understand it. This reward system is what motivates and interests the public in science. If there is not clear incentive, non-scientists aren't interested.

My family is a prime example of this. My mother prides herself on staying up to speed on what scientists have to say about newly discovered health information. Growing up, I distinctly remember her watching popular shows like "Dr. Oz", "The Doctors", and other various programs on the Heath Channel. She then would convey this information to me and my sister, insuring that our family stayed healthy, a clear reward. Members of my family, however, are not interested in other topics in science, such as climate change. Climate change has plenty of tangible rewards, but to some people, these rewards might be less obvious. Perhaps it's because health science directly impacts our daily lives. The more indirect sciences are viewed by the public as less important, and therefore create a knowledge gap. Upon interviewing UC Berkeley students, this idea was tested.

One science major and one non-science major were both asked to explain their attitude towards science. Not surprisingly, the science major had a greater grasp on the applicability of science. That is, with the student majoring in a science, there was a positive attitude and overall interest in science and its application. As for the non-science majored student, there was disinterest and slight confusion. This student had a harder time understanding the rewards of science stating that it lacked "clear, tangible results".

This slight knowledge gap exists even at a prestigious university such as UC Berkeley. This gap highlights and perpetuates the barrier that separates scientists and non-scientists. Non-scientists, the general public, don't understand the importance of science and cannot clearly see a benefit. It is possible that this disinterest stems from the innate selfishness of people. If the science is not relatable, not pertaining to health, then it's not important. In order to amend this barrier, scientists need to more effectively convey the importance of science. With this barrier intact, the public will continue to have disinterest in science and continue to fail to understand its benefit.

Khalid Al-Rayess: The Role Social Media Plays

Social media dominates most of our daily lives. However, the fashion in which it dominates our lives is very slanted. Popular social media sites such as Reddit, Facebook, and Twitter rely heavily on a mechanism that involves feedback. On Reddit the most popular articles get upvoted by users to the top of the site. On Twitter, trending articles get retweeted and if enough users retweet articles it becomes trending. On Facebook, articles reach the top of your newsfeed if an article reaches a lot of likes. Everyday this feedback mechanism takes place. But, this mechanism comes with costs and benefits that undermine the everyday user's inclination/interest in science. One of the many benefits in this feedback mechanism is the fact that users on these sites are able to draw attention to important issues that occur on that day. One can't escape their newsfeed or twitter feed without seeing the same article over and over again. This serves an important purpose in that it establishes to a user that they should learn about this issue and be held accountable for it. However, there are many costs that come with this mechanism. One of these costs of this mechanism is the fact that these topics that get liked or retweeted are usually in the same category. On a day to day basis, I usually see articles about astronomy, animals, or climate change. Rarely, do I see articles posted about the history of science, evolution, or discovery of new drugs. A reason why these topics such as astronomy or climate change keep trending is due to the fact that these topics are more appealing and interesting to daily users. By having an article be more appealing, it generates more traffic to the site that published the article. Thus, this slanted topic choice makes users become interested in only certain scientific topics that "hot" or more appealing. Futhermore, users on these sites who have little science backgrounds are only exposed to these certain types of scientific topics. Thus, this can create a false representation of the spectrum that science covers. As a daily user of social media myself, my perception of science was solely based on the articles I saw on a day to day basis on these sites; however, it wasn't until I was in college and took more specific science classes that I really grew new knowledge for science and a passion for it. Social media although has it benefits in getting certain topics publicized, users on these platforms need to reevaluate what is being shared. In a world, where millions of people are constantly on these sites 24/7 it is important for these sites to attain feedback for certain issues not just for attention, but rather for dialogue, engagement, or interest.

Xin (Natalie) Yu: Social Pressure

My personal interest on the project is to investigate some of the negative consequences that come from being not informed about science, and generate some reasons why the public would want to pursue science. Personally, I am a non-science major, therefore pure curiosity on science has never been strong. Due to the fact that science is becoming increasingly important in our society, I feel the need to know about science to be able to communicate and be respected. Even though I am enrolled in higher education, I still feel ignorant and behind when I do not keep up with current science trend, therefore I read science articles purposefully to inform myself about new science findings. This paper is my attempt to represent the majority of non-science related public. Learning about science for many can be seen as investing in social capital if the individuals themselves are not purely interested in science itself.

Because the lack of knowledge about science is not only immediately associated with ignorance, but also lack of intelligent conversation topics, lack of awareness, disconnectedness with the world, and lack of common sense. All of these factors leave one feeling vulnerable and inferior. In the social aspect, one is at a disadvantageous position in forming friendships with peers. The social component of understanding science is not to be neglected. As science becomes more predominant, the psychology of wanting to fit into the society is deeply embedded in one's mind. In order to have a sense of relatedness to the world and feel connected to other individuals, one of the many personal incentives to understand science is to satisfy the human desire to possess the feeling competence and relatedness. Our success cannot be completely personalized, and much of it is through our social connections and the opportunities we get from these connections. To get these connections, it is crucial that we know science and are able present ourselves as competent.

Making social connections is part of the process of becoming successful, which is defined here as having decent jobs high pay. (My colleague Mark Chang, talks in depth about the relationship between knowing science and career success) Nevertheless, science is important in cultivating an effective way of learning, because science consist of more than a never ending list of facts to memorize, and is a way of learning about the unknowns and reevaluating existing theories. It allows people to have alternative approaches and effective problem solving strategies. The society has grown a wide appreciation for this kind of thinking, with the emerging trend of start-up companies—scientific and innovative approach to tasks. Consequently, with a scientific mind gives one the chance to succeed and allows one to advance his or her place in society.

Our group has designed a survey to collect data on the hypothesis that non-science major students pursue science for social reasons more than the science-major student, and our data suggest the hypothesis to be true. We created a survey of six different incentives and we ask the participants to rank them from one to ten, one being the weakest and ten being the strongest. Non-science major receives an average score of 6.5 in the category of conversation topics while the science major receives a lower score of 5. The higher score of the non-science group in this category suggest that they want more social benefits out of knowing science than their science-peers.

Although learning about science is beneficial regardless, for the non-science public, the society has increased the demand for knowledge on science, many are pressured to learn about science to keep up with the growing trend. Almost as a requirement for success, as our values have changed with the growing importance of science, we psychologically want to pursue science to be able to fit in with the society.

Mark Chang: The Incentives Behind Public's Demand of Science
Thoughts before the project:

Science is humanity's greatest achievement, but it is sad to know that many of us are not able to share the joy of understanding science. Way too often, people feel excluded from the scientific group, and the more they feel excluded, the more they debate about science, in order to gain some knowledge about it. Every day, people use smart phones to shorten the distance between them and their friends, they drive automobiles to almost everywhere, they use home appliances constantly, but they do not understand the science behind those machines. They do not know what side effects there are behind the use of appliances, like the phishing websites that might steal one's personal information and that one is constantly monitored by the authority as long as he or she has a cell phone in hand. Assuming the existence of those people who do not have the best interest in their heart, general public gradually knows that the less science they know, the easier they could be fooled and manipulated by others. So in this case, science is much demanded by the public, to survive and to thrive in this world.

Besides the fact listed above, other reasons why public demand science include but not limit to: science enables public to imagine a better future in general since the more people know about science, the higher their salaries are, and the higher their life quality can be, science can contribute to our health and well-being, since science is a main facilitator for us to make right choices at the right time, and so on. Hopefully after this research on public's demand of science, I can better understand public's understanding of science in general.

Reflections after the project:

As the group member who compiles all the interviews, I got a chance to listen to all group members' interviews, and some striking things I got out of them are:

1. Health is a huge issue of non-science majors. Since they do not mess up with all variety of science, the particular fields of science they seem to concern the most are heavily related to normal life experiences.

2. Although not everyone is enjoying learning science, most people do realize the power of knowledge, and are willing to consume some amount of science at a low cost.

3. Early life experience in science really matters. In the interview, many interviewers had mentioned how a good/bad elementary teacher encourages/impedes a student to pursue a career in science. This is a serious topic we need to concern when we try to bridge the gap between the scientist and the public.

4. Science majors generally rate higher on how interest they are in science topic while non-science majors generally care more about the conversation topics they can generate from scientific knowledge and the job opportunities they may have. This discretion shows the incentive behind public's pursuit of science are both spiritual and physical.

http://mcbfifteen.tumblr.com/post/118679617750 Infographic

http://mcbfifteen.tumblr.com/post/118679840660 Audio

Part 5: On Becoming a Scientist from a Non-Science Background

Investigating different types of non-science backgrounds and the journey (including barriers) to becoming a scientist.

Emily Diaz:

As a first-generation, low-income, and historically underrepresented student, the two programs that have helped me develop my passion for pursuing a career in the sciences and establish a strong platform on which I balance my life at CAL have been the Educational Opportunity Program and the Biology Scholars Program. Considering those factors, my background has been the biggest barrier for me in becoming a scientist from a non-science background. I identify as coming from a non-science background because throughout my k-12 trajectory a learning environment in which students were encouraged to pursue science lacked. A lot of my k-12 teachers were not professionals in that field and therefore were not really able to stress the importance of science in one's everyday life. My family also had difficulty encouraging me to engage with science because of the lack of financial support and knowledge about science. On an academic and personal level, I noticed how much the public - in this case, anyone that could have had an academic influence on me such as my teachers and family – lack an understanding for science due to academic and financial disparities. However, now that I have transitioned into a higher education and am able to use my background to access resources that are specifically available to students like myself, I feel more inclined and motivated to pursue science. EOP and BSP have helped me navigate the resources I need to physically, mentally, and academically prepare myself to transition into a future a scientist at a higher education level.

The Educational Opportunity Program (EOP) is one of few programs on campus that is geared towards recruitment and retention of students like myself. The academic counselors within the program have holistically helped me find solutions to every problem I have encountered as a student. Financially, EOP has waived off my gym membership for the past three years because they know that is the main place where I take care of my health. They have offered me the Discretionary Grant that has funded medical expenses SHIP does not cover. The Discretionary Grant will also help me in alleviating some of the financial burdens in preparing for and attending medical school. Aside from that, EOP was first to offer me an internship and an employment opportunity where I have been able to develop gifts and talents that I have bottled up within me because of fear that I would be reprimanded for being my true character. EOP has kept me grounded and continues to help me learn how to balance my personal life with the rigorous demands of Berkeley that I was never prepared for.

The Biology Scholars Program (BSP) has also been another influential program that continues to keep me motivated in pursuing a science career. Because of BSP, I have been able to establish a small family and community outside of a big lecture class in which I am able to learn with other students that identify similarly to me. BSP provides mentorship, a vast amount of resources, and it has increased my circle of networks. This program has allowed me to feel comfortable not always knowing everything in the sciences, but while also giving me the courage to never underestimate what I do know. Because of BSP, I am continuously reassured that I will be a successful scientist and therefore my path in the sciences gets more and more defined as the years progress.

EOP and BSP are two programs that have helped me develop a sense of belonging and ownership in such an intimidating university that continues to attempt to weed out first-generation, low-income, and underrepresented students from following their passions.

Leilani Gutierrez-Palominos:

Professor Matsui is the leader of the course, "The Public Understanding of Science," in which he shares his wealth of knowledge and experiences. Despite not coming from a science background, Dr. Matsui is an extremely influential scientist who redefines what being a "scientist" means. To Matsui, being a scientist encompasses various experiences, including giving back to your community. Being a scientist is more than aloof figures in lab coats holding test tubes. He emphasizes that each of our private lives, goals, cultures, and experiences affects the way we become scientists and what we do as scientists. "This is not just about how well you do your science, but where you come from, where you go back to - all those influences outside of the laboratory beyond the university."

Growing up, Dr. Matsui's family and friends did not understand his love for science: "My mother, especially, didn't like the fact that I asked questions. She would call me a lawyer. 'You're always such a lawyer, you always have to know the answer to everything!' (laughs) and so I guess it was by nature that I was very inquisitive." Dr. Matsui extended this curiosity throughout his daily activities, such as when he worked in the fields and when he would get sick. When he worked under the hot sun as a field laborer, he would observe nature and inquire about farming techniques. In his early years, he also had allergies and asthma, which caused him to wonder how his body was reacting to external triggers. Dr. Matsui recognized his science interest came from within and, despite his limited resources, he would try to find solutions: "If there's a problem, I want to figure it out."

Dr. Matsui is both honest and humble and he believes a significant barrier in science is the discrepancy between what scientists currently are, and what they should be. Becoming a scientist from a non-science background means each person's experiences and hurdles are different and should be taken into account. The knowledge gap between a non-scientist and a scientist is large and can seem intimidating. I asked Dr. Matsui how he built the confidence to defy the odds and pursue science. Dr. Matsui's response was sincere and gave insight to what it is like to come from a non-science background: "I will make a confession...to this day, I doubt myself. And that's deep within. For those of us who come from the outside, we're always being tested. Always. It's called competency testing. Are you good enough? Do you belong here? Maybe there's been a mistake. Sometimes when I get into it with people with positions of power, I'll first have that old feeling of not trusting myself... then I'll say: No. I'm here. I'm at the table. I belong here. So I have to have that internal dialogue with myself. It's not like I've left my old self back there. We should never forget where we came from - but some of what we bring, doesn't always help us. It's important to be aware of our doubts, and having internal conversations as well as with a close group of people you can trust allows us to overcome those doubts. Nobody succeeds alone." Dr. Matsui believes in hard work and standing firm in what you believe in. Dr. Matsui carries the experiences of all the cultures he has been a part of and he tries to represent those who normally do not have a voice in science.

Gila Juárez: The First Generation Perspective

How and where someone is raised is a determining factor in how one's life progresses. We now see students coming into science with very little predisposition to the material. This is not to say that students are less likely to be successful, because that would not be the case. Focusing specifically on First-Generation students, I wanted to investigate how it is that someone with little to no exposure to science wishes to go into a STEM field. I opened it up to all STEM majors because I felt, from personal experience, that STEM is a community in itself that many first-generation students unite under because of the similar challenges they have faced to get to higher education.

I have been able to collect several responses to a survey asking about specific barriers that have been overcome or present throughout students' journey through STEM. Doing my best to generalize as little as possible, I have observed various patterns in the answers to critically important questions that include the following:

- How do you define "First-Generation"?
- Would you say you had a decent exposure to STEM in your community growing up?
- Are there any specific experiences before or after beginning study in your area of interest that impacted your journey into/through STEM?

From this survey, the general working definition used by survey-takers of a "first-generation" student would be an individual who has been raised in a country their parents immigrated to and/or they are the first of their family to attend an institution of higher education. It was not my intention to only gather responses from individuals who have been raised by immigrant parents, but I do think this rather consistent overlap to be very significant. It is representative of a great number of the individuals within communities of first-generation students. Speaking from research and personal experience, a significant number of this student population is also the first to graduate from high school. That in itself is so important to even begin thinking about a career in STEM and retention in institutions of higher education. It's not just about getting in, but also about staying to cross that stage at graduation. I feel that this tied my topic back into the chapter: lack of resources is one limiting element, or barrier.

The responses also include a steady answer of not having a decent exposure to STEM in their background, and yet all expressed a passionate interest in STEM. These interests seem to originate from similar experiences that placed quite a bit of emphasis on a steady presence of "disparities" among low-income communities. What I have consistently seen from the responses and their mention of lack of resources from a scientific standpoint, and other academia as well, and from my own experience, is that the main challenge has been an equity gap. Not so surprising—something else consistent has been a desire to close that equity gap through progress in the STEM field, and bringing what they can back to their communities, and other disadvantaged communities too.

How is this information representative to first-generation students across the spectrum? It really is not a generalization, but a pattern that I have noticed to be pretty common. These students from several different universities are motivated— almost all that participated in the survey are aiming for a doctoral or professional degree—and have faced challenges from early in their educational careers. Lack of resources and encouragement from educators is an unfortunate experience many first-gen, (and consequently low-income) students face. The bright side is then the fact that many of these students rise the occasion and hope to bridge the gap between their community and the science that they choose to pursue.

Overall, coming from a non-science background is not a complete disadvantage when pursuing a career as a scientist, and other STEM careers. It poses challenges, yes. Personally, it has been difficult to come from a disadvantaged background where many people don't understand your drive to pursue a scientific career when there are social injustices. But to be honest, there is a new kind of drive that pushes me to go farther. This investigation/narrative has presented me with the opportunity to relate with other students and to be able to bring some kind of awareness to this topic of barriers; that in itself fulfills my goal to raise scientific awareness quite a bit

Google Form Link "Why STEM?"

http://goo.gl/forms/8yvFkABJsR

Tarun Mendoza:

I would like to introduce this write up by saying that this project is based on personal experience. All my data and claims unless specifically cited are based on my experiences from childhood, growing up as a Hispanic/Indian in central California.

While I was growing up, I was surrounded primarily by Hispanic then Caucasian communities. The then average income was relatively low, and many of the Hispanic student were first generation (as was I). Most had accumulated disadvantages in education because their parents were immigrants and may not know the value of education. Even if the parents understand that education is important for the students, they generally were unable to aid the students to progress them further due to possible lack of knowledge. Whenever my friends (who were first generation) and I had questions about the material, we could never go to our parents because our parents didn't know the material. Some of us understood the value of education because our parents told us that it was needed to leave the current situation. However, those who didn't value education or were just frustrated with their lack of understanding of the subject would destroy resources (such as writing in books) and just not care.

However, afterschool programs and education based clubs aided many students in their quest for higher education. Even if the students at first didn't want to go to college, the students at least went to these clubs and programs for understandings of the current material and learn the basic material. Most of whom that I talked to said the main reason they went to the club was because they had learned the value of education and needed help getting to a point in which they can achieve their goals. These clubs and after school programs allowed these students to be even more qualified for college. However, these students had learned the value of education from somewhere.

So what causes these students to learn the value of education? One of my friends that I asked said, "My parents should me how hard their jobs and daily lives are. They even should me how hard it was to pay the bills will their jobs. This made me understand that I need to go to college to live comfortably. Even though I struggled a lot and my parents couldn't help me, I tried my best. These after school clubs like MESA and after school tutoring helped me get to the point I needed." This one student shows that parents can be one of the best avenues to the understanding of the value of education. Another student I interviewed said that he was motivated by a speaker that his school brought in to motivate student into the track to college education. However, even though these students were motivated by these methods, there are obviously other students that are not influenced by these methods. However, there are ways to maximize the number of students that would want to go to college for math or science. The main methods that every school should do are: First, try to have the parents explain to the students the struggles life has without proper education and the possible benefits math and science training will have. Second, school can try to stress how much after school programs/tutoring can help you in not only class now but also in the future to allow for a better education and life. Third, try to figure out what speakers might help the students lean to towards a higher education. Maybe scientists, engineers, etc. that have made a contribution to something that may have affected the students daily life.

After talking to others on the topic and discussing it with some group members, I learned some new valuable information. The main differences in everybody's lives are backgrounds and livelihood. This may seem obvious to some, but these two factors greatly aid in someone's choices and aspirations. I lived in a less wealthy rural neighborhood were most of my peers were first generations. As explained, in the previous paragraphs, I had explained how this posed challenges for all of us. However, I talked to some that were from wealthier areas. One of the people that I talked to lives in the bay area and was from a wealthier background. He said that the school that he went to would have a lot of academic clubs, many of which were scientific based. Not only did they have these clubs but also specific days that had speakers, science day, etc., which he said helped. Along with the school's influence, his parents would stress the fact that education is important. Parents such as his are already well off and understand that the education in their fields (generally math, science, tech, etc. related) was what propelled to that position. This is why more people need to put attention to education. The youth will not be able to understand its value by themselves. They need someone to show them. This is why more funding needs to put into schools. With this funding schools can do more events, get better equipment, etc. to help the process. Then with the students new knowledge about the value of education they can use that equipment to push towards their goals.

Regardless of what background someone is born into, everybody deserves the option and opportunity to know about the value of education and the possibility of going to college for the sciences. If every school did everything they can to not only educate the students and parents on what education can do to help students, then we would have more workers in all fields, especially in math, science, and technology.

Cristina Batarse: The Struggle

Many would not associate sociology with science; however they have more in common than one could imagine. In Sociology, there are these terms called "high track" and "low track," which ultimately depend on race and class (Oakes 65). High track is when an individual has been encouraged to think critically and creatively, whereas in the low track, it is more about memorizing and being obedient (Oakes 78). Based on the descriptions mentioned above, it is evident that the tracks try to gear people to certain careers or jobs. In high tracks, they are encouraged to pursue higher education and become extraordinary in whichever professional career they decide to pursue because they are given the necessary tools to succeed (Oakes 80). As for the low track, they are trained to be obedient and are given the necessary tools for labor jobs, and other jobs that do not entail much education (Oakes 88). But what happens when individuals from low tracks set higher goals for themselves than society had planned for them? These individuals face many challenges in getting to where they aspire to be. Individuals from low tracks who aspire to become scientists have a much harder path than those in high tracks who have the resources to excel. Those in low tracks come from non-science backgrounds, and majority of Biology Scholar Program (BSP) students also come from low tracks. Nonetheless, that has not discouraged any of the BSP students, who are first generation, minorities, and come from a low-income household, to pursue a scientific degree. Therefore, I interviewed three (3) BSP students regarding their transition from high school to a university such as Berkeley, as well as some of the obstacles they have had to encounter while pursuing their scientific career. This relates to the overall group's goal because I demonstrate some of the challenges and experiences that first generation college students undergo to achieve their goal of becoming scientists. My specific part of this chapter on "Barriers" is to shed more light on the barriers first generation students have to overcome to be at the same level as their peers who come from a science background.

The struggles that I am about to present are in no way a generalization of first generation, minority, or low-income students. I am simply demonstrating some of the challenges that certain individuals go through, which may or may not be relatable to the general public. From the three people that I interviewed, two are Latinos (a female and a male) who are from Southern California and the other individual is a Middle Eastern male who is from the Bay Area. All three individuals are minorities, first generation college students and come from low-income households. During the interview process, all three interviewees had similar responses in regards to their transition from their respective high schools to Cal. They had difficulty navigating the university system because they were naïve about what classes to take as freshmen, which classes they should or should not take together, and figuring out their financial situation. Moreover, the two Latinos also mentioned how the transition was even harder for them because they felt they were always being looked down upon because they are Latinos. Although these were some challenges that they had to overcome early on in their college career, the struggle still continues, as they try to overcome "self-doubt." They said that there are times when it is arduous for them to be confident because they feel that they are below average when compared to others at Cal who come from well off families who have been on the "high track."

Nonetheless, they said that becoming confident in their ability to be successful is a work in progress, as well as time management and self-care. For the Latina female and the Middle Eastern male, they acknowledged that their high school was definitely "low track." They said their high schools did not provide them with AP courses, which they felt set them back in terms of not being well-prepared as their college peers who had taken AP course since their freshman year in high school. Unlike these two individuals who knew that their high school was not providing them with the necessary resources to succeed in college and pursue a scientific degree, the Latino male believed he was being well-prepared. This individual graduated top of his senior class and felt well prepared for Berkeley; however, this confidence quickly dispersed once he began courses at Berkeley. He said that Berkeley humbled him and an example of this is when he met his formal roommate. His roommate had taken Calculus CD, which is far more advance than the course he had taken-- Calculus AB. At that point, he felt average and learned that he would have to work twice as hard as his roommate to get good grades. His mindset of him coming from a "high track" because his high school provided some AP course changed instantly when coming to Berkeley. He was now surrounded by the most elite students from schools all over the world.

Hence, individuals who come from "low tracks," have a more arduous journey when pursuing a scientific degree because they are not as exposed to the more advanced material that gives others a head start or the fundamentals to succeed in prerequisites for science majors. Despite the long journey that has been and will be ahead of them, all three individuals are determined to graduate with a scientific degree from Cal. They are attributing majority of their ongoing success to BSP because it has provided them with genuine academic and personal support. BSP is a program that aims to create a level plane field for low track students, so that they may be at the same level, academically, as those who were fortunate to have been placed at a "high track" for most of their life prior to college.

Work Cited

Oakes, Jeannie. 2005. Keeping Track: How Schools Structure Inequality. New Haven: Yale University Press, pgs. 61-92: oakes.pdf.

David Lamb: A Narrative Reflection of Different Resources

Income disparity has been increasing more and more within the past few decades. Research and studies have shown that parents who are from low-income housing and occupation with low wages tend to have children that will stay within that same income group. The poor will stay poor, and the rich will stay rich. These differences in wealth distribution plays a strong role on an individual's life choices. There are a lot of invisible barriers that can contribute to one's success and to me personally, the income disparity seems the most visible. At the bottom, you can really see the differences between the privilege and the nonprivileged; where the resources available might differ dramatically, having a completely different culture shaped by the socioeconomic background, even further contributing to the differences in opportunities, mindset of goals, and levels of understanding. Within this topic of public understanding of science, I will provide a narrative about the barriers and its effects toward one's individual development and understanding.

Looking back when I was just in 8th grade, I recall being voted "sleepiest" in the class. It's really interesting sometimes to think about it, to look at the trajectory I've paved myself now. At that time though, college wasn't an option available to me because I've never realized it as an opportunity back then, no one ever motivated or discussed with me about higher education. I would have never had the proper guidance or motivation to succeed and excel in high school if it weren't for certain outreach programs. These programs are patron to the success and education I have today, without them, I wouldn't have been the first one to attend a university from my family. I have 2 older sisters where only one of them had finished their associate's degree. It really makes me wonder if I'm a lucky one. Some outreach programs that are available don't necessarily just accept everybody, and those that are left out, are they going to be at a disadvantage when they're somewhat already at a disadvantage? How do these barriers, this difference in resources and opportunities have on individuals? Recalling further back to one of my 8th grade Spanish class, every day the students would be rowdy and disruptive, being ill toward and disrespecting the teacher and I'd still remember a few cases where she'd shed a few tears. It's just really absurd looking at it back then; where within this scenario, the learning environment seems counter-intuitive. "You are where you come from," an individual is reflective of towards their environment, it help shapes and nurtures you. Some of the students that might have been acting out in class were only motivated to because of the conditions that were made, where others also acted out; where nonetheless, affecting other students and their learning. From this discussion, I just wanted to hint a difference in the quality of education and structure; where although this is grade school, grade schools still contribute to your foundation in education and understanding.

If another was to comment on this story, they might describe it as if these "students don't intend to learn," but from my point of view, it's because "they don't know how to apply it and use it practically; a lot of them lack guidance and motivation." I know this because I might have been one of the individuals affected by the conditions, where these disruptions might have just demotivated me and somehow, made me gave up and just "sleep in class." The preparation for "college seems way too ahead for them to even be prepared for." Quoting this from the lecture with Dr Charis Thompson, where she mentions a girl that doesn't want to learn calculus because "she dislikes it." She made a point on how important it might for her and I find a lot of truth to that. A lot of the subjects you learn within earlier education helps lay your foundation. Personally I dislike math too because it's my worst subject, I am bad at calculus, but looking back, I actually liked it but might have somehow just lost touch with it, but maybe the truth is that I was just accepting the environment at that time, accepting the conditions that was set; where no one really cared about the education. That one math class in 8th grade, similar to my Spanish class was just as rowdy and disruptive… and from my point of view, these are sub-barriers to understanding, where the environment we live in might be shaped by our socioeconomic background and wealth; there's a lack of motivation to learn because a lot of the youths conceive the education as ill-pragmatic (where this ultimately serves as a different background, somewhat disadvantaged compared to others). Where we can also note the differences in education, and to further note, education is now being more and more expensive. The gap of understanding will further continue to widen as income disparity and resources further widen. This all contributes to the public understanding (of knowledge, not just science). There's this one quote where people say "Knowledge is power," so therefore if the public doesn't really have a certain understanding or knowledge, are we therefore powerless? There isn't just a need for action to aid in the public understanding of science, but there are also a need to help youths, especially those who are unmotivated and are coming from a background that doesn't inspire or set the tone them to be properly guided and to succeed. Some say education within grade school isn't important, but for me, from my experience and now looking back, it really does seem to lay your foundation. I recall in high school how my peers and friends would mention that I am "really smart," but personally, I was just "trying hard."

I am grateful to the programs that had reached out to me, but am wary about the differences in privileges and opportunities and the trajectory we have towards understanding. It might be harder for youths from underprivileged background to excel; where a lot of these youths might contribute to what we perceive as the public understanding, where also interestingly within the income group, we'll represent the majority or "the 99% in a comical sense," but I guess it's not really funny. Recalling back to Matsui's lecture, what's most interesting about the talk on public understanding, is the talk on change agents, such as Steve Jobs campaign with Apple "to think different," to become "change agents." What's more interesting is how I'm a beneficiary of Job's legacy, where one of the core outreach programs that helped and guided me was founded by Laurene Powell Jobs. These change agents serve as the mediator to the differences in opportunities and resources; leveling the playing field, making the space for understanding. Even college now too, where many individuals are at a disadvantage as we all come from different background, some resources out there like BSP, also serve as a space for understanding and support to the disadvantaged. Within the public understanding, creating the space for understanding can help bridge the gaps.

Siso Phouthavong: My Journey into the Sciences as a Pomo Indian

I would like to preface this write up by saying that I am speaking from personal experience and observations. My data and claims, unless specifically cited, are based on my experience growing up in the Native American communities of Northern California, primarily using my mother as resource for reference.

Growing up, I noticed that I lived a life very different than my peers in my low income, underrepresented locale. This could be due to the realization that my Native American community was a minority even among other underrepresented groups. I do consider myself a part of a marginalized community because historically and contemporarily Native Americans have been treated as a second thought in many regards, which has created innumerable barriers to education, and specifically the sciences.

Historically, land that was once very rich with culture was partitioned in such a way that the natives who initially roamed the land were forced onto small reservations, whereas the government was free to exploit the rest. At least in my family tribe, this had very startling consequences as my family was promised decent living conditions by the US government, but instead my family was left to fend for itself completely. More recently, natives have been sitting on the American back burning in various situations. One in particular is illustrated by the fact that natives are continuously and consistently the most underrepresented minority in higher education. Based on articles and data I've read, universities like UC Berkeley have been struggling to keep up their already low native population of students having origins in North America or the Pacific Islands, whereas other minorities are seeing much better outcomes. This stems from Native Americans not being exposed early on to opportunities and a lack of outreach programs that could potentially mentor young natives to help them realize their educational goals.

Growing up, I've noticed the main support to be from the family front rather than a collection of many individuals striving for a cause. The family culture in Native

American society is very strong, but the effects are differential, which is strongly dependent on each family, distance, and time. For example, my mother grew up in Crescent City, CA, a town that nearly borders Oregon. In addition, she was raised on the very isolate periphery of this city, which added to the already strong effects of distance. She grew up in a very traditional setting, and as a result, had a lot of family who she could turn to. In fact, any day she would expect to see any of her dozens of aunts, uncles, and cousins. They would aggregate and partake in traditional activities. This strong tie allowed my mother to find a home in nearly any one of our family. However, this tribal life style also had its draw back, too.

My mother, although provided family stability in the long term, she was unable to pursue and kind of educational goals-- in fact, she claimed she couldn't even imagine furthering her education even past the middle school level. These pressures against schooling were derived from the source which provided her so much support: her family. Despite these sentiments against education, she was always in the top of her class, and never had any trouble obtaining the highest grades in school. Her teachers were astounded, and often encouraged her to further her education in school programs, but these lacked the necessities she was so familiar with within her family. Indeed, it was the lack of resources for Natives in these educational outreach programs that halted her aspirations. This was common within my family, and because of this, she made the decision to end this cycle by pushing her children to pursue higher education, a privilege she was never afforded.

The effects of this, however, were not complete because my siblings all had difficulties trying to balance the tribal culture and educational path. School was often second nature for them, but due to family pressures, they internalized these feelings which had devastating effects. Perhaps, if there were resources available that could balance the Native lifestyle as well as our satiation for education, Natives would be better mobilized in society and the sciences. However, this is difficult because many of my family still have barriers that are preventing them from realizing their true potentials.

Speaking from experience, my tribe has had troubles visualizing college. Out of my 350 member tribe, less than 10 of them have went through college successfully. With rates as low as my tribes, it's seen as virtually inaccessible to go on to college. In many ways I feel it took the push and support of my entire tribe to send just me onto college. Although this may be seen like breaking through the barrier, there seems to be more ahead. I'm the first in my immediate family to go to college, and I'm also the first in my tribe who decided to pursue science, a field that is very rigorous, and at UC Berkeley known to weed out underprepared students.

So, what is the underrepresented-underrepresented minority to do? In an age when there are many initiatives to help underrepresented minorities gain a voice and fighting chance in the public arena, why are Native Americans still facing barriers concerning these same initiatives of education-- the bridge that facilitates the movement of opportunities? Perhaps this is my chance to step in. As I said before, I had my entire family and most of tribe to support me in reaching my educational goals, specifically those related to science. Although some of the science I am learning is not necessarily important for tribal culture, it does bring about new views, and with that comes new opportunities. Although this is unusual in Native communities, my bridge to science was my family; however, for many other Natives, this bridge is very hard to visualize because their family's priorities lay elsewhere. Perhaps the key for Natives to successfully enter the sciences is within people like my mother, who eventually pass this key down to other people like myself. Now that I am in possession of this key, I realize that I can have a great effect on younger generations of Natives. People who can inspire have a great responsibility to pass on their knowledge and sense of motivation. I desire to pass my key of knowledge and motivation down to other Natives so they too can be enlightened by sciences.

CHAPTER 2

PUBLIC PERSPECTIVES IN SCIENCE

Even the most mindful of us may struggle to make scientifically informed decisions.

Chapter 2: Public Experiences and Perspectives in Science

By Vin Y. Lay

Edited by Julie Mendoza

Science holds a role in the imagination of the public and its professionals. Americans take pride in the accomplishments of their researchers and the scientific discoveries made within their fields. Still, there continues to be a disconnection. A disparity in awareness despite public education in science reveals deeper issues that impact our effectiveness as a community. We are considered one of the global leaders from medicine to food, environment to technology, yet we carry opposing views as a public. The perceptions of science and scientists vary when compared among different socioeconomic classes. This separation tends to relate to differences in education. It is well known that investing in scientific advancement is a priority for government and that this is enforced by public support. If disparities in understanding continue to grow it may result in the regression of our scientific fields.

One of the goals of this publication is to bring attention to the gap as it exists today. Recognizing topics that are lost in a socioeconomic translation such as vaccination, cloning, or artificial intelligence, may bring us to make unified decisions. In the midst of an internet culture being exposed to new information is easier than it's ever been. Finding ways to filter the accurate information from the inaccurate has become more difficult. It becomes a question of how well equipped students are after their public education and whether the insight provided is enough to have good judgment within an environment constantly campaigning different truths.

Deciphering how individuals perceive their own gap can lead us to the tools that will build the bridge. This chapter covers experiences that reveal the differing realities between communities. We will look at the sources of scientific information amongst different socioeconomic classes and other cultural divides.

Most Americans tend to obtain their scientific pallet through school from K-12. A smaller percentage continue to develop this pallet through higher education such as the bachelor, master or doctoral levels. One issue that we address is that America is a country of high immigrant populations with a public science education offered only in English. The matriculation tends to forget the special circumstances of many of its students. UC Berkeley student, Minh-Thy Nguyen, says that her mother "craved the sciences, wanting the knowledge of the field, yet was unable to completely understand the terms in English." Another student, Paolo Joaquin, describes in his essay an angle of the Filipino community's perspective. He finds that because of the "diet, language...and culture" even those that attend school see little relevancy of the subject in their daily lives.

A second matter discussed in this chapter is how age impacts how concepts of science are retained throughout child development. Rather than focusing on the biological factors in learning we question how sociality contribute to a learning experience and ways to engage various circumstances. For example, UC Berkeley student, Keana Richmond, recognizes that "scientists that come from families outside of science have the unique opportunity of being in close proximity of 'the public'". Having learned abstract explanations in school while surrounded by those without the same understanding, these individuals are in the position to deliver simpler and direct translations of research. UC Berkeley students Diana Bahena and Sophie Ballard mention that a key source of science related material for many is social media. It is used as an outlet used to communicate across all levels of understanding. However the current of true information within these outlets is often countered by what is equally convincing but untrue. There is not yet a way of mediating the overwhelming amounts of data making many individuals only source of science utterly unreliable. We will continue discussing this issue further in the chapter.

Examining the gap of perception in its context will allow us to have a stronger grasp on the way the general public digests science at different ages, socioeconomic class, and communities. In addition to asking how knowledge is gained we hope to discover how it is used, as well as what status, history, or demographic influence these answers.

Part 1: Changing Perspectives with Development

Investigate the way that the general public perceives science and gains their scientific information throughout the stages of their lives.

Sophie Ballard:

As people develop throughout their lives, their views of science change. These views are influenced by many different factors, such as family, friends, past education, and media. At later stages in life, many adults and elderly are no longer in school, and, thus, do not have the same exposure to scientific information that they did in their younger years. Therefore, during this portion of their lifetime, their scientific knowledge must stem from somewhere else. Their knowledge may be based on what they learned in school many years earlier, what they read in the newspaper or social media webpages, or what they see on their television when they are eating dinner.

The working-class adults spend their time working hard to make enough money to survive and provide for their families. The elderly may be dealing with the physical or psychological issues surrounding the aging process. Despite this, it is still vital that both of these demographics, as voting citizens, have access and exposure to current scientific work. In order to investigate this further, I interviewed several working-class adults in my own family. When asked where they received their scientific knowledge, they answered that it came from their college and high school educations, discussions with their friends and family, and from popular science websites such as Yahoo News. They suggested that, to integrate more accurate science into their daily lives, the media work more closely with scientists to incorporate current scientific knowledge into the media. This would help expand the individual working adult's knowledge, and it would also help spread that information to others via their social sphere on topics such as vaccines for children. One single working mother proposed that the focus should be on making current science more interesting and applicable to non-scientific adults by finding ways to relate it to the general public's interest and by eliminating unnecessary technical jargon. Considering that the daily lives of working-class adults are filled with working for a living, scientists should work more closely with the media to incorporate modern scientific knowledge into popular television programs, social media, and magazines in order to reach them.

Additionally, scientists should facilitate good scientific education in the elderly, a portion of the population whose time is likely consumed with dealing with issues related to aging. A senior citizen that I interviewed made an excellent point, which was that the best way to reach this age group would be to offer science classes designed for the elderly. In order to do this, science programs should be implemented in nursing homes and at senior centers so that the elderly can keep their minds active in their remaining years and can expand their knowledge if they choose to do so. Additionally, another senior citizen I interviewed suggested that accredited universities and research institutions should create television programs. These television programs would present current research being conducted in various scientific fields so that even the elderly who are bedridden, and thus cannot attend senior citizen science courses, can still access the scientific information in a medium that is enjoyable.

Research institutions and the government should incentivize scientists via grants, tenure, raises, and recognition to spread their research to the general public. They could do this through working with the media and teaching courses to the elderly. Spreading current scientific information to the voting population, regardless of age, in a way that works well for each age group is critically important.

Jacob Cota:

What is Synthetic Dance-ology? We, synthetic biologists, use Synthetic Dance-ology as a way to educate individuals about synthetic biology by engaging them in a fun activity like dancing. Our last engagement event was at the California Academy of Sciences, where we told our participants what synthetic biologists do, about engineering bacteria to kill a tumor, and the steps that are in that process. As Synthetic Dance-ologists, we use role playing cards with specific functions for bacteria and tumors, and have dance moves associated with each function. Throughout the dance, we tell the participants the specifics of the scientific process in a digestible way. Then at the end of the dance, we summarize the overall process of tumor-killing bacteria, and offer to explain more or answer any dying questions they have about science and/or synthetic biology in general. Included at the end of this article is an excerpt of what we tell participants during our Synthetic Dance-ology introduction.

When first introduced to Synthetic Dance-ology, I was skeptical of people's interest and the effectiveness of teaching non-science oriented individuals. However, after experiencing and helping out with this process, I found out that our method worked effectively. People (ages ranged from 21-40) appeared really interested in not only dancing but learning about synthetic biology. They asked a lot of questions during and after the dance and even interacted with some of us synthetic biologists after the dance. They asked questions about tumors and their formation, and about treatments and how they worked. An individual asked, "why do we use bacteria to target tumors instead of viruses or some other chemical in pill form?" Some even asked about our field of study in the labs we are apart of. For instance, one individual asked what I research in my lab, and how that relates to the general public.

I believe Synthetic Dance-ology is a success! I feel that getting that level of engagement and enthusiasm from our participants was promising. It led me to believe that one form of teaching science in an effective way is through engaging your audience with something they can relate to and have fun with, while including science in the mix. The individuals that participated in Synthetic Dance-ology came to the California Academy of Sciences to learn about science. This shows that museums are a great platform to educate people about science related topics. Individuals of the scientific community should use this opportunity to try new creative ways to educate non-science oriented individuals. Synthetic Dance-ology takes people, college age and older, and draws them in with dance while surreptitiously teaching them basic synthetic biology concepts. Afterwards, when the scientific ice has been broke, the floor is open to questions where any budding scientific mind can ask questions of experts in the scientific community. Many individuals that came to the California Academy of Sciences, are a particular subset of people who know that science is approachable or have a developing interest and seek scientific knowledge to quench their thirst. Synthetic Dance-ology is a perfect platform to satisfy the scientific curiosity of these individuals. Furthermore, those individuals who view science as inaccessible and intimidating will find that this is not the case with the help of Synthetic Dance-ology.

EXCERPT: We're here to help you learn about synthetic biology. So what exactly is synthetic biology? Synthetic biology is an exciting new field that combines chemistry, biology, computer science, and engineering. Synthetic biologists take parts, or genes coded by a DNA sequence that have specific functions, and string them together to engineer a final product in a biological organism. But we have a mission. We Synthetic Dance-ologists, like synthetic biologists, have selected some functions from other organisms in nature that can give you, the bacterium, the power to *detect and destroy the tumor hiding in this room.* We are going to insert the instructions for these powers into your DNA by biological engineering similar to synthetic biologists. For more details, you can speak to one of the scientists at the tables. For now, you can imagine it similar to wanting a superpower, like flying, and us taking a gene found in nature, like those from a pigeon, and inserting it into your DNA... of course, it's a bit more complicated than that. Your bacteria superpowers are going to be taught to you now through movement. These products are meant to have a tremendous positive impact on people's lives. Some examples include the development of an anti-malarial drug that can be produced by bacterial cells. Similarly, insulin can be produced more cost-effectively. Synthetic biologists are also engineering biology to generate new biofuels and to produce biodegradable plastics.

Today, we are going to be Synthetic Dance-ologists and enact a tumor-killing bacterium project that was first started by researchers at UC Berkeley and UCSF. Imagine that a tumor is inside of Amanda's body and it needs to be eradicated to make her healthy again. Instead of cancer treatments that can affect all of the cells in Amanda's body like traditional chemotherapy, we want to be smart synthetic biologists and engineer a bacterium that can find the tumor and destroy it without harming any other part of Amanda. As Synthetic Dance-ologist instructors, we're going to teach you the parts needed to do this successfully through instructions for single movements. Then we'll put those movements or parts all together into a single host bacterium to generate a final product, the tumor-killing bacterium. We'll also learn how to identify a tumor by its movement.

Betsy Rosales Avalos:

It is in early school years that children develop their interest and abilities, so it is a critical time period to expose them to in order to instill a sense of affinity and curiosity for a subject that they may be discouraged from pursuing due to social or religious factors. According to the Community Resource for Science, "The digital divide is augmenting the opportunity gap in science, as students from low income and minority communities often lack access to science enrichment outside of the school day." Many kids lack an environment outside of school in which they observe and experiment with the natural and physical world; but, many also lack education in the classroom setting due to a lack of resources and a lack of professional preparation among school teachers to teach systematic studies of the structure and behavior of the world. A survey mentioned on the CRS website says a reported 80 percent of 923 Bay Area elementary school teachers said "they spent less than an hour each week teaching science, including 16% of the total number who spent no time at all on science." Children are said to be natural scientists: they are extremely curious and love to ask questions. Thus, it is important to introduce them to a subject they have a fascination with at theses early stages because it allows them to think critically about the world around them.

I volunteer with CRS, specifically with their Bay Area Scientists in Schools program (BASIS). BASIS is a "science role model volunteer program" that targets these specific schools to provide students with a basic foundation of the behavior and structure of their surroundings through in class presentations including a lecture on a topic with easy to understand information and in class experiments. My BASIS team and I visited a classroom in Berkeley recently to teach on colors/light and their basic components. Interestingly enough, the kids always had answers to our questions such as "What are the primary colors? Secondary? And what colors compose these secondary colors?" As well as answering questions on what happens when we shine a white light on a prism. When it came time for the experiment, which included color chromatography and mixing colored paint, the kids loved being hands on and insisted on experimenting further with different colors. At this age, kids are learning about the world and how things work; thus, have a natural fascination with the world that seems to wither away with age due to various factors. However, exposing all kids to a subject of their natural interest science allows them the opportunity to grow with it and develop their critical thinking skills useful to other subjects in school as well as everyday life.

Moreover, an intriguing aspect of this presentation was the fact that we taught it bilingually in a class of diverse 1st graders. Being exposed to science in a different language encourages abstract thinking because one is presented with the opportunity to think and respond differently.

Some of the students we reach out to would otherwise lack information on real life science issues in school nor have gain inquiry-based learning experiences. Furthermore, those who miss out on science education in their early years are minority and low income, which may be a factor in explaining the lack people of color in the science fields. Kids love being independent and touchy, so increasing an experimental and observational education in elementary schools is critical to diversify the field of science. The fact that two latinos and two latinas presented in a bilingual manner allows them to expand their viewpoint on who can be a scientist by subtly deconstructing gender and racial boundaries. Programs similar to BASIS are important to have and to nourish in order to bridge the gap between the people and their perspectives on natural and physical occurrences in our world.

Keana Richmond:

Although perceptions of science typically form during the developmental stages of a person's life, they continue to change throughout our adult lives as exposure to science occurs in lots of different ways. The quality of our education, our parents' views, and our access to experiences all influence the way in which we think about everything - including science. There is a "sub-group" of adults within the public that has a unique kind of access to science: those whose children are scientists but who are not involved with science themselves. Parents suddenly gain access to a wealth of information that may not have been available or relevant during their personal and intellectual development.

Educating the youngest generation has proven numerous times to be overwhelmingly effective at spreading awareness of some of the most pressing issues that the public is either unaware of or unfamiliar with. Examples of these efforts include projects and outreach programs that were implemented in schools on the island of Guam where I was born. A few of these projects were geared towards spreading awareness of the state of the coral reefs that surround the island in elementary, middle, and high schools. Kids were taken to beaches to learn simple biological concepts and to observe the corals first-hand. Beach clean ups and basic surveys of the near-shore reefs were carried out by the students - many of whom brought their parents along. Surprisingly, the parents began reaching out to the facilitators of these programs to learn more about what they could do be more involved with the work. Another program that aims to promote watershed management in order to improve water quality for near-shore reefs and fisheries engages community participation by educating children about the importance of these issues. One of the most prominent impacts of this program was that adults in the community were pressured by their kids into adopting more sustainable practices. More than 1,000 community members went on field trips to see the watersheds and engaged in management activities, contributing more than 2,000 hours of volunteer work to the program. Before these programs were introduced into schools, these adults were either unaware of or indifferent to the issues that Guam's environment currently faces.

In my own experience, influencing a change in perspective of a parent as a student in the sciences becomes imperative as the depth with which I can engage them increases with my age and level of education. My mother is an artist and self-proclaimed nonscientist. When I tell her that I'm learning about ocean circulation or the latest study on GMOs, I find myself having to emphasize the purposes and implications of those issues. I can't just show her the numbers on a graph; I have to connect my education to applications in the familiar world. The significance of these issues is lost when we only try to communicate *our* way.

Exposure to science can be beneficial to not only the people receiving the education, but to the people closest to them as well. A person's familial relationship with someone can often have a much more effective impact than an interaction with a stranger. Scientists that come from families outside of science have the unique opportunity of being in close proximity with members of "the public." It is of paramount importance to communicate science to the public because science has implications. A data point is just a number on a line on a graph. It is the responsibility of those that have the opportunity and possess the skill-set to communicate the information on that graph because it affects everyone, not just those that can read it. Understanding the meaning, consequences, and solutions of global issues is what scientists and nonscientists alike must do.

Diane Chong:

When one uses the words "science" and "religion" in one sentence, their connotations of being inherently antagonistic towards each other are enough to make people cringe. Indeed, society has established a curious relationship between the two subjects that has its rather antagonistic roots in the 17th century Enlightenment period. However, the modern world has learned to take an interdisciplinary approach in many fields spurred on by a movement by secondary educational institutions across the country. Many institutions have begun to encourage students to master concepts in both the sciences and the humanities through implementing multiple core curriculum requirement. For example, the University of California, Berkeley requires mastery of English, mathematics, American culture, foreign languages, among other subjects from their students for graduation candidacy that teach both raw scientific material and facts as well as respect for different ethnicities, cultures, and religions. Likewise, individual student pioneers seem to have taken up this call and many from both the scientific and religious community seem to be increasingly revolutionizing the way in which the world perceives these two subjects through their personal syntheses and mindsets.

One common stereotype of individuals of faith that can be witnessed quite easily if one ventures onto Internet articles on religion or science is that they are somehow less educated in the sciences in order to believe in their religion. In order to investigate this stereotype, I created a Google survey and distributed it to my general Facebook newsfeed and to my Christian fellowship group, specifically requesting, aiming, and filtering out the responses to current college students. All of the responses came from 34 students attending the University of California, Berkeley and the University of California, Los Angeles. This survey asked for students' majors to gauge how familiar they were with the sciences, their religion, any recognition of their mastery of the material such as a prestigious summer internship or post-graduation job offer, where they received their science news, what their opinion was on science and religion, and a specific example in which they could use to explain and substantiate their opinion in the previous question. The results from the survey revealed that 75% of the students interviewed were both religious and majoring in a technical science, dispelling the myth of uninformed religious individuals. Many of the students majoring in the technical sciences reported that they continued to gain scientific knowledge and perception of science through their classes and online scientific articles, and for students majoring in the humanities, many turned to science articles and referred back to their primary education and any college science courses for their scientific knowledge.

The results from the question regarding science and religion yielded interesting qualitative data. The effect of religion on these students' perception of science seems to not have affected most of their performance in the class, the rigor of their schedules, or their firm grasp of the material taught. This was concluded from the lineup of jobs offered to graduating seniors of various science majors such as chemical engineering, molecular and cell biology, and biochemistry perhaps the ultimate authority on validated mastery of the science courses we can measure up to this point. The effect of religion, then, seems to have only affected how they perceived the current unknown in science. Many of the religious students noted how they attributed the mysteries of science that has eluded scientists for years to a higher being. For example, a Presbyterian Christian UC Berkeley senior majoring in Molecular and Cell Biology explained how some biological molecules don't act the way they should, and how chemical atoms combined into molecules act differently from the way one would have imagined based on how they function independently. She was one of the survey participants who attributed these details to the handiwork of a higher being. Another Presbyterian Christian UC Berkeley 3rd year student majoring in Integrative Biology questioned the mutually exclusive nature of the Big Bang and the creation of the world by a God, and questioned why both couldn't exist and how one could have used the other to create the world we live in. However, none of the survey participants expressed that these mysteries were not worth investigating and the value of discovering the unknown was not diminished simply because they attributed the intricacies to a God. For example, one Presbyterian Christian UCLA junior majoring in Chemical Engineering summed it up in the following manner: "Science is real and visible all around us. Nothing about it is 'fake'. Our understanding of all things, however, is not perfect and this is precisely where the interplay between science and religion begins. Science and religion thus cannot be inherently antagonistic."

    Though religion and science have historically had antagonistic connotations with each other, the increasingly interdisciplinary world allows for a change in thought and chance at synthesis. The students surveyed are all living proof that highly educated people continue to merge their understanding of the world with the objective understanding of the world. All the students interviewed concluded that their knowledge of science was impermeable, and religion only clarified their perception of science by providing an explanation for the unknown. It is important to note that this did not reduce the importance they placed on discovering new scientific governances of the universe. To these students, science and religion worked in tandem to contribute to their understanding of the world, a mentality both professors, religious, and nonreligious students alike would do well to consider and respect in order to create a more harmonious, progressive, and interdisciplinary world.

Diana Bahena:

The study of the physical and natural world is a very important aspect of society. It is the root of many discoveries and advancements. We observe it implemented in various facets of our lives, from the curriculum of schools to social media. Nevertheless, people have a different comprehension of the physical and natural world due to their age and their various experiences. For example, middle school students are highly influenced by the education system to the extent that most of their understanding of the physical and natural world is based on the information given in classes.

As I interviewed my fourteen-year-old brother along with his four other friends, I discovered that their natural science classes had sparked an interest in this topic. However, they were all interested in different concentrations of the science field. When I asked my brother Francisco his favorite topic, he thoroughly explained that he loved learning about atoms and their different properties. As his eyes widened with excitement, he mentioned hydrogen bonds, ionic bonds, and covalent bonds along with their definitions. I then preceded to ask him if he would ever consider in the future being a scientist or working in the field of natural science. He responded with excitement, "Yes! I want to do experiments like Bill Nye." Through this comment, I came to the realization that science teachers in middle school still frequently show the program, *Bill Nye the Science Guy*. As I interviewed my brother's friends, they also mentioned these videos and its positive impact on them. They explained that it reinforced the material that teachers taught and helped them understand topics that were very difficult.

Consequently through this interview, I realized that educational videos have played an important role in my brother's scientific knowledge. This is because the videos of *Bill Nye the Science Guy* spark an interest in the physical and natural world by being both informative and entertaining. As Bill Nye is explaining and giving out facts, there are constant sound effects and graphics. His voice also changes throughout the entire video, but always with a sense of excitement. Thus, this has pushed my brother to be constantly engaged throughout the entire time, allowing him to retain the information that Bill Nye is stating throughout the video.

As a result, media has played a key role in informing my younger, middle school brother about the physical and natural world. At this stage in his life, he is very curious and by having educational videos that are very dynamic and noteworthy, his interest in the natural sciences has increased. Moreover, these videos present the field of natural science in a fun manner and therefore, it has caused him to want to potentially become a scientist in the future. Hence, I believe educational videos such as *Bill Nye the Science Guy* could potentially help other middle school students in increasing their scientific knowledge and curiosity towards the field of natural science.

Vanessa Sarmiento:

When I was small, I wanted to be a ballerina. I didn't want to be a teacher and I thought doctors had the most disgusting job in the world. I though science was difficult, confusing, and involved too much math for my liking. As I grew up, I slowly began to realize that becoming a ballerina wasn't very feasible. I still had no interest in becoming a doctor or doing anything in health and medicine, but when I took my first biology class in seventh grade, I realized I really liked knowing how babies were made and that meant genetically and physically. I became interested in science and medicine simply because our biology teacher taught us about genetics in an interactive way, by using an activity where we had to determine what genetics our "child" would come out with by chance. This activity was one of my fondest memories of my first biology class and it helped me realize that was has truly sparked my interest in science is seeing it in action. In elementary school, there were several times where on field trips, we would get to see demonstrations and at times even participate in things such as dissections, engineering demos, and other interactive exhibits. These opportunities were very prominent throughout my elementary school career and to me, it made science and what scientists do a lot more tangible and exciting. I wanted to participate more and more in these programs but as I grew up there were less and less opportunities being offered. Looking back on this now, there were not less programs being offered to me as I grew up, but the structure of them changed into something I no longer found interesting. Instead of dissections I was offered internships and instead of robotics seminars I was offered summer research programs. This shift in the programs makes sense if you look at it in the perspective of an adult; if a child is growing and learning more details and technical subjects, they should be given a similar amount of detail and technicality within their offered scientific programs, but this did not catch my attention.

As high school students, teenagers are often faced with varying levels of stress and responsibility. Students are expected to go to school and do well as they prepare themselves for the next step in their lives whether it be pursuing higher education or going straight into the workforce. High school students are faced with the challenge of meeting adult expectations while still being treated as children whenever it is convenient. With all of this in mind, it is difficult to imagine how a student can be newly exposed to the scientific world. We put a large emphasis on science in today's society because we are a world that has come so far due to our technological and medical advancements. We rely on technology on a daily basis and when we are sick, we rely on the medicine that has been engineered by the great scientists before us. Within this world we are encouraged to pursue careers in STEM fields but we only make an effort to really sell these fields to younger children. This effort to "sell" the sciences should continue on especially into high school because at this stage, teenagers are beginning to make life decisions that will affect their career choices for the rest of their lives. As students are exposed to science throughout their lives, one problem that is seen is retention of these students' interest in the scientific field. We are expecting high school students to remain interested in a huge part of our evolving world by testing them on facts and figures and not really showing them how that relates to the greater picture in today's society.

An ideology that can be effective in retention of interest for high school students is one that is already being used for this age group. Instead of giving out exams and expecting teenagers to memorize everything we tell them, we should be teaching them to think critically and we should also be helping them connect the science in their textbooks to the science in their bodies and cell phones. A program that uses this ideology is one called Peer Health Exchange. Peer Health Exchange is a non-profit organization designed to help teach sexual and health education to ninth graders in underserved communities. Often times, high schools in lower income areas are unable to provide adequate health education due to lack of funding and this is where Peer Health Exchange comes into play. By utilizing a workforce of college volunteers, Peer Health Exchange (PHE) is able to go into classrooms and teach students about different health topics such as drugs, STDs, and how to deal with health related decisions. What makes this program stand out is that it does not focus on simply lecturing to students but instead, PHE focuses on demonstrations and interactive activities that push students to think critically about what answers are right for themselves and in certain situations. This environment garners much more enthusiasm and participation because the entire discussion is based upon how much the students contribute and ask questions. A similar format could be extremely beneficial to raising interest in the scientific fields.

Through my experience working with high school students through Peer Health Exchange, I can see that there is a positive response to complicated topics if students are not pressured to memorize all the facts and details behind the subject. Whenever a lecture becomes a discussion, topics tend to become more interesting and memorable for students. Topics such as biology, computer science, and chemistry can all easily be turned into either extremely confusing and tedious or fascinating and inspiring subjects. Students are continuously developing and their interests are always changing so we should similarly continue to re-inspire them to develop their interests as they mature into adults; though we shouldn't take their development and growing maturity to mean that they no longer need interactive activities. Most people tend to respond better to presentations that are stimulating through either discussion or active participation and high school student are no different. Applying this ideology does not mean that we should stop teaching students typical scientific subjects in detail and through textbooks, but we should be able to supplement these lessons in order to help them see the "bigger picture" that most texts fail to provide. By maintaining the same level of interactivity and enthusiasm towards recruiting students into scientific fields throughout all stages of education and perhaps even into adulthood, we can help students truly grasp the large role that biology, chemistry, computer science, and various similar fields play in our world.

Part 2: Science, Family, and Community

Looking at science from the family and community perspective.

Mar Jean:

Being multicultural, I have had the opportunity to look at science from different lenses. Meaning, I have had also the opportunity to explore science in both of my cultures, differently. For example, growing up in a Mexican family and around a Latino community I got to see in what ways science topics were exposed to us. At first, I had a hard time coming up with anything because I could not find ways in which science was exposed to us. However, as I visited my family over vacation, I notice that my family has certain channels they watch. Univision and Telemundo, which are two of the biggest Spanish speaking channels Latino families watch to see news, sports, soap operas, and there's even something for children on the weekend mornings. What I found more interesting is that science was being portrayed in these channels in ways that I never realize. I always saw these channels as entertainment channels but recently I have seen great changes that involve individuals of the Latino community.

Univision, for example, has infomercials conducted by Jorge Ramos, a journalist that is well respected in the Latino community for his involvement with the community and always informing us about the latest news. Unlike most infomercials that try to sell products, some of the infomercial that Univision offers inform the public about different issues in the Education, Health, and Technology fields. Ramos takes about five minutes presenting to the community how to get involved or dismissed myths the public may have concerns on. For example, there is an infomercial, that Univision conducted, where they talked about the measles outbreak and how the vaccine is not link to down syndrome, a concern many families had. Speaking to my family members, especially my older cousins, I learned that when the measles outbreak accrued they were concerned since they did not have much information. I interviewed Fernanda Perez, my cousin, and a young mother that arrived to the USA 10 years ago. Perez states, " Pues la verda no tenia mucha informacion sobre la vacuna del sarampion y si mi hija ya la tenia o no". Perez explains her concerns stating that she did not have much information and was not even sure if her daughter already had the vaccine or not. I then asked her if she had notice that Univision started to offer infomercial that talk about health, education, and technology. Perez, "Si alveces los miro y son muy interesantes como lo traves estaba uno que explica los desventajas de no dormir mucho pero la verda cuando salen commercials es mi oportunidad para ir al baño o hacer algo antes que empiese mi novella de nuevo (laughs)". Here Perez is explaining how she has watch the infomercials and gives an example of one she watch that explains the disadvantages of not getting enough sleep, however; she does mentioned that she does not always watch them because they come during commercials and usually this is the time she runs to the bathroom or does something she needs to do before her soap opera starts. Her mother that was listening to our conversations also agree with many of Perez's points. Some of the positive affects of the infomercials are that it does reach Latino families and they tend to be informative. However, one of the negative affects is that is only played during the commercials, a time many people find as a break to do other things while they wait for their show to resume, thus miss out on the infomercials.

By Univision providing infomercial that point out concerns individuals may have, for example, they also did one regarding measles vaccinations. Helps out the Latino community because it is reaching out to many communities across the country that gets in tune with this channel. Even though right now is just an infomercial hopefully in the future they can create a program where is more than 5 minutes and thus create more awareness and engages the public to speak to one another about controversial topics and most importantly it brings science to the living rooms. Looking at science from my family perspective and being close to the Latino community I also get input from them and get to see science from their perspective as well as my own.

Wendi Ruiz:

Growing up in a low-income community and family, I was never exposed to science or any aspect of it. Unfortunately, I am not the only one. Many families from low-income communities do not obtain a high school diploma nor did they attend college. Therefore, many families and communities aren't exposed to the concept of research/science because we aren't correctly informed about it or we don't have the resources or knowledge to learn about it. Our group's goal for this chapter is to get different "public perspectives" about science from different people in different communities and understand why they portray science that way. This final project is on how low-income communities view science in order to make scientists aware of how certain communities portray science in hopes that they make science more accessible and understanding to the communities.

I interviewed 5 people from low-income communities, who wish to remain anonymous for this project, and asked them how they felt about science and how their families felt about science. One out of five students responded that his parents did attend college and they viewed science as something informative and helpful. Four out of five students had parents who did not attend college and weren't exposed to any type of science and they claimed that their families were clueless to what is going on in the science community. When I asked them why, they simply said, "our parents aren't exposed to scientific material and if they are, they wouldn't understand it because of its daunting terminology." On the other hand, there were many families that didn't know about science very well, but knew some things about it due to "pop-science" and science related channels/movies, which is more common for the public. When I asked the students how they view science, I received different responses. All five students emphasized the importance of science and how it should be something everyone has access too: something everyone understands. When I asked them why, one student said, "it is important for everyone to be aware of what's going on in the science community, but it is also important for scientists to address the public properly." Finally, I asked how can we make this possible, and the students thought about this question for a few minutes. A student then conveyed that scientists could go to these low-income communities and do the same thing we are doing: asking us (the public) how we view science, and then they could "take this knowledge and use it as a source to promote science more in these communities."

As a solution: we should make science easily accessible and understandable to the public. Once we see how different communities and families view science and what they think about, it will be clear to scientists/researchers how we view science and will therefore, do what it takes to "bridge the gap" between the community and science.

Nancy Cuevas:

An individual's perspective of science does not emerge simultaneously without external factors. Indeed, a community and family's experience shape what someone thinks of science. Therefore, my final project is focused on how Latino farm workers perceive the effects of science in agriculture. My plan consists in investigating how new agricultural methods have affected individuals working in the fields and what these people think of science. This topic emerged after taking a Latin American Studies class that exposed the "development" of agriculture in Latin America in the late 20th and 21st century, where chemicals and other scientific initiatives have greatly affected people who work with for example Monsanto, and my mom's own experience in the fields. Therefore, I am very interested in analyzing how traditional farming was replaced with GMOs as a technique of neoliberalism, which implanted Westernized knowledge and consequently generated many health ramifications upon communities working with such practices.

Indeed, I examine how marginalized communities perceive science; 90% of the people I have interviewed do not even know about the effects of science: the chemicals, pesticides, etcetera. My mom, for example, did notice all the pesticides being used in the fruit she worked with but never questioned the effects of such chemicals. This derives from her lack of knowledge/education and the employment dynamics since inquiring would probably make her lose her job and put her immigrant status on jeopardy. When I asked her whether she saw an airplane flinging pesticides or chemical to the fruit she worked with, she told me that she observed them regularly, but that she never questioned what they contained and even to this day, she still does not know what they were or had. I was inspired by her answers and decided to further my project by conducting interviews of people who have worked or are still employed in agriculture. This included family members, friend's parents and other close individuals. Jose, the father of my UC Berkeley friend was narrating the story of his brother, who died while working in the fields. Despite never attaining a clear answer of the causes of his death, Jose articulated that "Yo culpo a los quimicos con los que trabajaba. Él era la persona que trabajaba todos los días regando las plantas con unos líquidos que jamás supimos lo que contenían o los efectos de tal." After hearing the stories of people who have been affected or worked with such chemicals, I noticed that the lack of information about such science had to do with the corporatist's interests, the worker's lack of education, and the farmers' deficiency of the official language, English. Four out of the five people I interviewed had no strong perspective about the new science being used on agriculture because they entailed no knowledge about it. None of the interviewed subjects knew how to read or speak English and none of them had a formal education higher than middle school. Thus, I concluded that Latino farm workers' perspective of science in agriculture is very limited due to their lack of knowledge of chemicals and pesticides. Without knowledge about science, chemicals, pesticides, GMOs, and Monsanto, most of them consequently did not know of the possible health ramifications that these involved.

My project relates to the overall group's goal to analyze how science is perceived from a family or community lens. Therefore, I desire to serve as a third culture between the public, this time being Latino farm workers, and science. My specific part of this chapter on "Science, family and community" is to look at the big picture of the effects of science on Latino communities and agricultural workers' perception of science.

Rafael Mejia:

Science perceived in film and television is often seen as a mad man who is conducting experiments surrounded with multiple glassware filled with different colored liquids and smoke coming out. This type of "science" could be one of the earliest forms of introduction to the idea of science and what a scientist does. This reference to science can be perceived as a stereotype. According to the Oxford Dictionary a stereotype is a "widely held but fixed and oversimplified image or idea of a particular type of person or thing". Science can be portrayed either negative or positive depending on certain social circumstances and level of education of individuals. Families who lack the level of a college education, influenced by cultural circumstances and are part of certain socioeconomic status have a narrow idea of what science really is, outside of the typical Frankenstein science. In order to investigate the topic, science stereotyped within families, this paper will focus on the gap created within certain cultures and levels of education.

First and foremost science is usually generalized like I stated before. According to Marcelo Gleiser of the National Public Radio, it has to do with "invisibility". In order to understand science, one has to know someone in real life that does science. This is one way to avoid over generalizations within the public domain. Marcelo Gleiser suggests for "researchers and their graduate students […] regularly visit public and private schools" (Gleiser). Which is great to improve science for the future. However in certain familial cultures the topic of being a scientist and doing science already exists a certain image.

I held four different conversations to help distinguish how different families typically view the idea of science and someone being a scientist. The first two conversations were conducted with members of my immediate Mexican family. The third was the mother, African-American, of a friend I shall call "M" who is studying at UC Berkeley and lastly the Father and Mother of one my closest friend "A" from high school whose family is Vietnamese.

Mother, late 40's, homemaker, Low Income, Education: Mexico - Grade School

Q: What do you think when you hear the word "Science"

"I think of a man wearing a lab coat, conducting experiments"

Q: Have you ever done something scientific? Like in school?

"I remember when I was in grade school we conducted experiments"

Q: What kind of experiments?

"AH! I do not know, we just followed the instructions based off the teacher and a textbook with an image of a man holding glass with smoke coming out"

Aunt, late 40's, Pre-K Teacher, Middle Income, Education: California - Community College

Q: What do you think when you hear the word "Science"

"I think of people conducting experiments, discovering new things."

Q: What kind of things and have you ever done something scientific? Like in school?

"Like how to prevent deadly viruses from spreading, and yes, I remember in school we conducted an experiment with baking soda in a bottle blowing up a balloon... that's science right?"

"Yes, that can be considered scientific."

<u>M's Mother, early 50's, Unemployed, Low Income, Education – US High School</u>

Q: What does your mom think when you tell her you're a science major?

"They think I'm a genius and that I'm going to become a doctor and be rich. But when I tell them I want to do other types of science they don't understand. Like they don't get that there are other science fields outside of being a doctor."

Q: Like your major huh?

"Yeah! I love animals and see how they interact with each other is so interesting!"

<u>A's Mother (Computer Security) and Father (Accountant), Both Mid 50's, Middle Income, Education – UC Berkeley College Graduates</u>

Q: What do your parents think of when they hear the word "science"?

"My dad find the topics of science extremely dope, he finds scientific discoveries fascinating. Then again my dad is super geeky. My mom does not care too much about the subject."

Q: They both have a solid understanding of science?

"Yeah, in terms of research and scientific discoveries."

Conducted from the conversations I had with the people close to me. An image of what science already is to them comes into play. With my mother she immediately thought of science as someone wearing a lab coat conducting crazy experiments, or science was thought as something related to the health field such as with my Aunt's reasoning and "E's" mother lack of understanding what science really is. And lastly "A's" parents were both college graduates who were familiar with the subject of science because the father sparked an interest in it. This is only a sample of families that all have a different understanding of science, usually ending up being over generalized. Whether the families were exposed to science or not, a gap exists just from the small sample of conversation. Science is expanded from the idea of man in a lab coat, to being a doctor and finding new discoveries in research. The gap exists and has been influenced by the level of education these individuals have had, therefore science is stereotyped depending on the different levels of education. But is there a true set definition on what science really is? Sure science can involve a man wearing a lab coat, science is involved with the health field and sure that science is involved with discovery. There are just so many levels to determine what science really is. To close this gap I like Marcelo Gleiser suggestion of having professors, researchers, and graduate students to visit schools and expose the meaning of science at an earlier start of education.

Adela Ramos:

When I hear the words science, family, and community used in the same context, ideas of culture, and perspective come to mind. This also involves traditions and the way that generation after generation has been raised. Throughout the time I lived at home, I think there was the general consensus that we would follow our traditions and religious views as we always have. For example, the idea of the morality of abortions was an interesting topic to talk about and even evolution. When I started questioning religion in the context of evolution, my sisters and I would get into discussions about our convictions. I started questioning my convictions around the time I was in middle school, since we were actively learning about science and I was very interested in that. I feel that, at least in Mexican culture, there is a gap between openly discussing science-related topics and actively seeking out information. This might arise from how culture and religion are highly intertwined. With this in mind, I would like to explore the implications that stem cell research has on Mexican communities. This topic is one that I have only recently become more familiar with and although this area of research seems very promising in determining human health and longevity, I would like to take a look into how religious people may perceive this. We have been talking about bioethics and "morality" came up so I thought about how perhaps many Mexican families may not be in favor of studying stem cells and changing their function for fear of "playing God". I don't consider myself a very religious person and because of that I don't think and feel the same way that my family members may feel about certain controversial topics. Overall, I think it is more challenging to present science-related research to people who have some sort of religious background, although not impossible if they are willing to listen. I mostly base my assumptions and convictions on what I know and understand about the topic in question. I intend to use my own family as a case study. My goal is to present how people with a Mexican culture background perceive topics such as studying stem cells (including embryonic stem cells) or even gene editing. Research done on stem cells implies that much light is being shed on how to control cell development and growth and I don't think traditional Mexican families would be too comfortable with the idea of being able to "control" or alter our cell's growth and development.

Upon asking my mom what her thoughts were on *science*, she told me that science is a subject that requires a lot of knowledge and training to understand. I realized that maybe my question was not very valid since science is such a broad and large subject; it is a large umbrella term that involves a lot of other subjects. I tried to then narrow my question down by asking her what she thought about researchers and doctors having the power to control what babies can look like physically. She said that this was not OK since God has already predisposed us to be the way we are and that we should not mess with anything since it would not be normal. I asked her what about the case in which doctors could get rid of diseases before the child was even born and again she said that would not be OK. She said that family should always take care of family and that we should never reject a family member just because they are different. This in turn made me think about the way we have been raised and how since we were young my siblings and I were told to love each other unconditionally. My mom also told me that it was wrong to make fun or not accept anyone because of a medical condition. She spoke of that in terms of how God would certainly punish us if we did not accept his will. That reminded me of the term "Catholic guilt" and how Catholic people are raised to believe that they are born sinners and therefore must always repent for their sins. Sure we can always just choose to do the right thing but there is also an inherent fear that if we don't follow God's will, we will be punished. I asked my friend (21 years old), who was raised in a Catholic family and he said he saw no harm in stem cell research being done if it is done for the right reason; in order to find cures to diseases that would otherwise cause harm to a person's health and longevity.

Minh-Thy Nguyen:

Culture plays a role in every person's upbringing: from the people who raise them, the community that surrounds them, and the lessons they are taught. As a result, culture becomes a prevalent factor in the path a person may choose, and whether or not it is inclusive of science. Growing up in a family with parents who had emigrated from Vietnam to the United States, Vietnamese culture was a force that guided the direction in which I chose to take. With a heavy emphasis on education, my parents instilled values of hard work and success in my brothers and me, in hopes that we would surpass what they had been able to give to us. Thus, the pursuit of science came naturally to me: it posed a challenge, as well as a pathway towards respect and stability that aligned so well with the beliefs that had been instilled in me. However, my perspective of science differs very much from my parents. Giving birth to me in the United States, my parents gave me the opportunity to easily follow a pathway involving science; however it was not as easy for them. My mother has always told me how much she wished her English had been better, or less heavily accented, so that she could have pursued another career in the United States. She too craved the sciences, wanting the knowledge of the field, yet was unable to completely understand the terms in English. In addition, studying science is not the most direct course to making money, requiring a lot of further education. In order to put food on the table and care for three children, my mother chose a track that would allow for her to work immediately after graduation, rather than one that interested her intellectually. Even so, she still loves reading popular science articles in the Vietnamese newspaper and learning as much as she can. Unfortunately, my mother also takes everything she reads in these articles to be fact, coming home to renounce certain foods or items because an article told her they lead to cancer, diseases, etc. Without a formal understanding of science, my mother is easily convinced by self-titled experts in the field and becomes swayed by studies that are not necessarily proven. However, unlike my mother's easy acceptance of scientific notions, much of the public tends to fervently disagree with scientists. Dan Kahan, (2010) talks about the notion of a "cultural cognition of risk" in which people form perceptions of risk that align or are similar to their own values and beliefs. This theory states that people try to protect their own beliefs and those of people similar to them in order to avoid conflict and protect social status (Kahan, 2010). Thus, people will outright reject certain experts, and seek others that will support their ideals. This theory also explains my mom's acceptance of science: her culture has placed trust in any person with a doctorate degree, and as she was unable to achieve the same, she believes they would not mislead her. Because the culture each of us is raised with becomes ingrained into each of our minds, we seek to protect it. If someone were to disagree or attack our beliefs, it would cause for question to arise about our entire upbringing, something many people are unable to accept. I am fortunate to be raised in a culture that encourages education and science, as well as to live in a time period in which I do not have the same worries as my parents did. Because of the struggles my parents endured, I am free to study topics that interest me, and can help clear any misunderstandings they may have in the realm of science.

Maria Rivera:

An important branch of science that has been emerging in the most recent years is Neurobiology, which is also my intended field of study and the topic that I will be focusing for the purpose of this essay in order to model a collaborative effort between a scientific group and a (mostly) non-scientific group to understand each other's perspectives when it comes to science. The scientific group in this case will be modeled by me as I ask various members in my family who do not work with science or who have a limited amount of knowledge regarding science, (and therefore represent a small sample of the public outside of science), what they believe Neurobiology to be; after which I will record their response when I inform them from a scientists' perspective what Neurobiology really encompasses.

When I asked my 21 year-old cousin who is a junior at Cal State Long Beach what her understanding of Neuroscience involved, she immediately mentioned that her first thoughts upon hearing the word were brains and neurons. Then, because she is studying psychology, she revealed her knowledge about the close relationship between neuroscience and psychology in regards to human beings and their behaviors, but that she did not know much about the scientific aspect of neurobiology or what else it may be applied to. I pointed out to her that neurobiology was also a collaboration between chemistry, computer science, engineering, linguistics, mathematics, medicine (including neurology), genetics, philosophy, physics, and psychology. She was immensely surprised to learn about the interdisciplinary nature of neurobiology and the fact that one who studies neurobiology must also learn about chemistry and physics on top of biology in order to fully understand the workings of the brain. She was also astonished by the fact that science itself did not seem like a different bubble separate from the humanities bubble when looked at in the context of neurobiology; the fact that neurobiology seems to bring both science and non-scientific fields of study together helped her see science in a new light, one that does not automatically regard science as a difficult subject that should be avoided at all costs but rather one that is also interesting due to its complexity.

"La gente asocia la palabra neurología con patologías muy graves o discapacidades, y por eso el mero hecho de escuchar la palabra neurólogo causa un gran temor e inquietud en las personas". When I asked my mom what her associations with neuroscience were, her response was to say that hearing the word itself caused a great fear in people and patients because one automatically thinks about neurologists dealing with pathological brain issues, tumors, and serious disabilities. Although it was a great shock, I did understand why for a patient to be referenced to a neurologist would be a cause for alarm on their part and can cause them to worry to such a great extent since neurologists are not known for dealing with everyday sicknesses such as the common cold. However, it is a common misconception for people to think that Neuroscience is only the study of cerebral tumors, hemorrhages, brain clots, etc. because it also deals with more common issues such as memory losses, concussions, or even dizziness. So in response, I let my mom know that not all visits to a neurologist need be for drastic circumstances since they can simply be to stunt the advancement of real diseases in our bodies. I let my mom know: *"Los motivos de*

consulta con neurólogos más frecuentes y en la mayoría de las ocasiones son las pérdidas de memoria no patológicas o los mareos inespecíficos. Por esa razón, es importante saber que no siempre hay que temer a la ciencia; en realidad nos puede ayudar a tener vidas más saludable y de mejor calidad".

It was important for me to see my family members' perspective on this branch of science and to be able to share a different view with them as well; when we put this form of interaction on a larger scale with the rest of our communities, we will finally be able to work together in understanding science as critical to our lives in a more appreciative light.

Marycon Jiro:

Science is thriving in the Philippines. More and more people each year are delving into engineering, health, and other science professions. The subject of science is implemented in the educational curriculum, and this early exposure leads to interest in science as a career option. However, most of the people that pursue the sciences are wealthy or well off. Many factors prevent the poorer Pilipino people from pursuing science-related professions. Unless they are extremely talented or have family members that were also in the sciences, family obligations, religion, and/or lack of resources, money and means override their passion for their science of choice. When I went back to my birthplace recently in the Philippines, poverty was almost everywhere in the midst of a progressing city. I believe that many of the poorer people could not have had sciences further from their minds when they constantly worry about feeding themselves and their family members at home. How can they think of science or invest money and time into a science related profession, two things that these vocations heavily demand, when they themselves do not have money or time? To bridge the gap between these extremely poor people and science in general, we must first address the poverty and lack of resources these people face. The Philippine government should address these concerns by helping their people better care for themselves and their family, and then funding the buildings to be involved in research.

Additionally, the quality of primary and secondary education exponentially worsens the farther away from the city, into the rural communities. While there is a big difference between public schools and private schools here in the United States, the contrast between the public and private schools is immensely pronounced in the Philippines. All I hear are stories from my family members, because I never attended public institutions in the Philippines. My family members back home in the Philippines are very lucky and blessed because most, if not all, of my extended family are educated, well-off, and supportive of the sciences. We live five minutes from the city, and my cousins are all attending universities back home, pursuing professions such as nursing, engineering, and architecture. They received a private education before college, a privilege that I did not even think about before coming to Berkeley. When I initially wanted to have my project be about contrasting views on science within my family members, I could not do it because despite my family's religious, Catholic background, they believed that, if the knowledge of the sciences is in good hands, society can benefit and progress, especially in health, education, construction, etc. In the words of my mom in my Hiligaynon language, a dialect in the Philippines, "*Dapat kabalo sang manga tao mag science para mag abanse ng Pilipinas*," or, "people in the Philippines should know science so that the Philippines can advance." Instead of constricting myself to just my family member, I looked at the Philippines as a whole.

Something that can help bridge the gap between poor Pilipinos and science is by allowing for more spaces in the Philippines for those who are educated in the sciences so that they in turn can teach and inspire their fellow people. Many educated Pilipinos leave the Philippines because they cannot find well paying jobs in which their area of study applies to, especially in research jobs. This leaves to a heavy "Brain Drain" where the educated masses of the Philippine population migrate out of the country. Also, those that can teach science choose either to teach in the wealthier parts of the Philippines or to not teach in the Philippines at all because there is not a lot of money allocated for teachers. If the Philippine government can offer additional stipends to teachers who choose to teach in rural areas of the Philippines, something that Ricardos Lagos implemented in Chile for his country's teachers, it would give the teachers incentives to reach out to poorer countries. While many people think that people should teach in the rural areas out of the kindness of their hearts, it is realistically impossible in the Philippines because money is already strapped.

Science is seen as the future by many Pilipinos, but most cannot take part because their present situations are dire. If their needs are addressed, discussions about science will be more widespread and not just concentrated to an elite few in the Philippines.

Part 3: Bridging the Gap

Expand our avenues when it comes to communicating science to different communities we like to call "the public."

Nicholas Garcia:

Throughout this semester we've been discussing how to understand the vastness of science and its topics. Even more so, we've been looking at how to communicate these scientific findings and issues to an audience that might be unaware. Science has relevance in society at large and it cannot be kept in this "ivory tower." We've called this issue "bridging the gap." This idea of bridging is not easy; in fact it continues to be the scientific community day in and day out. For example, one prominent issue in science is the use of pesticides. Pesticides control weed, insect, fungus, and other pests from crops allowing for cheaper food and clothing. However, there are negative health effects to farmer workers that work directly with such chemicals. These health risks are overlooked by the larger community because only that small aforementioned community. At the same time, we must find a way to expand awareness to the harmful side of science.

To begin the discussion of pesticides we look at the most used one in California, which is glyphosate. While glyphosate kills weeds and unwanted plants, it also possesses cancerous after-effects as well as non-cancer effects like diabetes, reproductive and developmental issues, and even Parkinson's disease. These by-products put these farm workers at risk for an unhealthy lifestyle but communicating to them these risks might be more difficult. On the other end of the spectrum we have consumers who are not directly impacted by these pesticides but whose consumption increases the demand for agricultural products and the use of pesticides. Therefore, they're harming farm workers too without even knowing it! So how do we communicate to two very different groups? Groups at which come from very different backgrounds, generate very different incomes, and hold very different priorities and values when it comes to lifestyle. After going through this course and hearing a variety of guest lecturers, it all comes down to, "Who is your audience?" When you pinpoint your perception of the public then it would be easier to accommodate to those backgrounds, and lifestyle choices mentioned above. For example, farm workers are in the low income bracket and maybe do not have access to forms of social media. So maybe we communicate through print or word of mouth. Conversely, consumers have that luxury to own a Facebook or a television set. The best way to reach out to them would be through those mediums. Nevertheless, when communicating these issues about pesticides or science at large, it's best to present it as if there is an impending danger so that way there is a call to action.

To me, reach out to the "public" that way is fine; however, it might not be as effective long-term. I believe that education systems should bring to light these controversial issues and start showing students that there is this delicate balance when it comes to science. Teaching kids at a young age could impact the long-term mentality of students and possibly inspire a kid to make a change.

In the end, these are all suggestions. Some might work; other might not work. Communicating these issues can only go so far when it comes to making a change in the science community. But in terms of bridging the gap, becoming aware of the issues is a great way to start empathizing and understanding the culture of science. Overall there is this big question of "How do we communicate these issues to many people (and different people for that matter)?" This idea to me is the basis of bridging the gap.

Gabriela Wantah:

At UC Berkeley, students are granted the privilege to major in a wide array of subjects. From Dutch Studies to Hebrew and Structural Engineering, Berkeley offers its students a variety of majors to choose from. That being said, however, while Cal is consistently ranked one of the top science schools in the world, I have noticed that the majority of the Indonesian students at Cal tend to shy away from majors, and even classes regarding the sciences. This leads to a large gap within the Indonesian community in understanding science. In my final project for this class, I interviewed 4 different individuals and asked them either why it is that they are majoring in a science, or why it is that they are not. I also asked them why they think that given the choice and resources, why Indonesians do not take more science courses or even choose to major in a science.

I, myself, was born in Jakarta, Indonesia, but moved to the United States when I was 3. Unlike the majority of the Indonesians on campus, I grew up in Los Angeles,

CA, and was thus accustomed to a very "American" style of living and learning. I attend Berkeley with the mindset that I will live and try to find work in the United States after I graduate. The Indonesian friends that I interviewed are all international students and plan to either work in the United States for a few years before permanently moving back to Indonesia, or immediately return to Indonesia upon graduating. I interviewed two science majors (one Psychology major, and the other, a Nutritional Science major), and two non-science majors (one, a Business Administration major, and the other, Civil Engineering). For the purpose of this paper, I will follow the definition of a "scientist" provided by the dean of Boston University. In the article her wrote, he states that while "scientists explore the natural world and show us how and why it is as it is," "Engineers innovate solutions to real-world challenges in society" (Lutchen 5). Engineers, therefore, are not scientists.

Every year at UC Berkeley, BISA, Berkeley's Indonesian Student Association has a membership of about 130 Indonesian students, 95% of whom are international students. However every year, there are only about 5 to 10 students out of those 130 who major in something science-related. The more popular majors include Economics, Statistics, and the many branches of engineering like Industrial Engineering or Chemical Engineering. When I interviewed my two friends and asked them why they were majoring in Psychology and Nutritional Sciences, they responded that their majors were very interesting to them and very relevant in their lives. With the information that they learned in their classes, they were able to apply it to their daily lives. They found value in what they learned and felt that they would continue to use that information even after they graduated. While Lily, the friend who was majoring in psychology, did like science, she enjoyed more of the social aspect of psychology rather than the scientific.

This year, there are roughly only about 5 students in the whole Indonesian community who are majoring in a science, and all of them were female. Of those 5 female Indonesian students, I interviewed two. When I asked them both why they had chosen the majors they did and not a science-related major, they both responded by saying that in order to practice medicine back in Indonesia, you would need a medical license and the only way to obtain one is to study medicine or biology in Indonesia and graduate from a school there. While that is so, Indonesia is not as technologically advanced as many other countries yet, so often times, students back in Indonesia lack the proper technology to learn or practice. My friends had chosen to major in Business Administration and Civil Engineering because with degrees in those majors, they would more easily be able to find jobs back home and elsewhere in the world.

At the end of the interview, I asked my friends to further elaborate on their explanations for why the statistics for Indonesian students studying science are as they are. Some of the reasons they gave me included the fact that science tends to "not be as flexible or applicable to real life situations" as other majors. Also, "there's more money in business" and because Indonesia is not as advanced technologically as other countries, "even engineers are reluctant to major in engineering" in fear that they won't be able to find a job in their fields after graduation. Not to mention, many of Indonesia's most common diseases, like Dengue Fever, are not known to doctors and professionals studying in America, so their education in America would be irrelevant to the diseases back home in Indonesia.

Due to the reasons stated above, my Indonesian friends at UC Berkeley are reluctant to invest time to study subjects like Molecular Cell Biology, Integrative Biology, or even Public Health in fear that their education would not be useful in the long run. And it is because of this that many of them study subjects like engineering or economics, and are not interested in anything science-related. My friends, like many others, only take biology-related courses to fulfill their breadths, and end up not retaining much of the information they learn in those courses because they were not enjoyable to them. According to Einstein's quote, "education is what remains after one has forgotten everything he learned in school," my friends are not getting a proper education in the sciences and as a result, are expanding the gap even further. In order to bridge the gap, students like my friends need to be shown that science is relevant in all aspects of life and that it does not just pertain to one study. The sciences are neither as intimidating as one thinks, nor are they only for those who want to practice medicine. By expanding one's understanding of science, students will be more willing to try science courses, enjoy them, and possibly end up pursuing their studies in the sciences even further, thus slowly bridging the gap.

Rattana Sot:

One of the major forefronts of bridging the gap between scientists and the public is through the means of education. Though education funding in the United States exceeds that of our national defense, there still is a major fault in our education system.

Why are some students better prepared for higher education than others? This education inequality is prevalent in the nearby city of Oakland, California. The dropout rate in Oakland K12 education is relatively high compared to other surrounding school districts: 21.7% compared to Berkeley Unified's 11.6%, Alameda City Unified's 8.4%, and Piedmont Unified's less than 10%. Graduation rates in Oakland are also relatively lower compared to nearby school districts, and the percentage of students who are accepted into 4year universities are also significantly lower according to education data from kidsdata.org. The close proximity of these higher performing school districts raise an alarm as to why Oakland Unified School District is struggling and other districts are not. One major reason is the teacher turnover rate. According to an article covering

Oakland's education system, "... close to a fifth of them leave the district every year."

With an influx of inexperienced teachers from programs like Teach for America, coming into a teaching environment where students have to deal with issues far more pressing than academics, students and teachers both are not getting the best out of what the education system has to offer. In fact, they are getting the worst. According to the same article, "Despite working in one of the least affordable housing markets in the country, teachers in the Oakland Unified School District have gotten only a few percent raise in the past decade. They're currently the lowest paid public school teachers in Alameda

County"(Dalmas). The district that needs the most funding in Alameda County is the district with the least funding. As a result, students in Oakland are subject to an education that is far from competent. They are subject to an education system that does not prepare them for higher education. Without an education that is fair and equal, the gap between the public, communities in Oakland in specific, and scientists is widened.

In fact, it is almost as if the communities in Oakland are left out of the the general public population. To combat this, one simple solution is to give students in Oakland the right to the education they deserve. This means providing AP classes, having healthy breakfast and lunch programs, hiring teachers who plan to stay long term and who can accommodate the needs of their students, having afterschool programs that tend to underperforming youth, etc.. Though this issue will not be solved right away, it is a step in the right direction. With more educated and motivated students coming out of cities like Oakland, the gap becomes smaller. Youth will be able to understand the sciences and bring back what they learn to their families. It becomes a chain effect where people are learn and the gap shrinks overtime. Though there are many ways that can close the gap between scientists and the public, implementing a fair and equal education system that takes care of all students should be a first step.

Johannah Perez:

Though religious exemptions to vaccinations have existed since the 1800s, the factors and mechanisms that inform the desire for religious exemptions have not necessarily been well understood. When people state that they or others they know oppose vaccines because vaccines are "against" a religion, there seems to be an implication that religious exemptions are strictly a result of theological objections. Often, religious reasons to decline immunization reflect concerns about vaccine safety or personal beliefs, rather than theologically based objections (Grabenstein, 2012). It is imperative to understand the sources of false information about vaccines, as well as the specific objections that religious communities make with regard to vaccines, before proposing solutions that can mend the gap between the scientific and religious communities.

The composition and preparation of vaccines are concerns that affect multiple religious communities. Hinduism and Jainism, for example, advocate non-violence and respect for life—even embracing vegetarianism as a means to respect all life forms—and thus may find issue with trace animal components in vaccines, as well as the killing of microorganisms necessary for vaccines (Grabenstein, 2012). However, Hindus and Jains also attest that violence in self-defense is justifiable. As physicians begin to understand the objections of these communities, perhaps more targeted and specific explanations about the preventative nature of disease can, in turn, explain the need for the killing of microorganisms. This argument can also appeal to the right of "justifiable self defense." A report about the parent perceptions of vaccinations indicated that "most parents with concerns ended up immunizing after having discussions with their physician about why the vaccine was important for their child" (Frederickson et. al, 2004). The physician, then, can play the unique role of a mediator that facilitates an understanding about the benefits of vaccination. Instead of disregarding religious objections as illegitimate, physicians should have a sensitivity to and acknowledgement of religious objections in order to effectively communicate with patients and, at the very least, offer another perspective uncolored by religious bias. The report (2004) acknowledges that "[f]amily physicians, pediatricians, and nurses need to be more fully prepared to have productive discussions with parents who resist or refuse immunization."

Indeed, the public shares similar concerns about vaccinations and vaccine safety, and knowing these general public concerns may possibly contribute to further understanding of the concerns of religious communities. The side effects of vaccines, for example, is a key issue that worries the public, regardless of whether they subscribe to a religion. In 1999, concerns were raised about a relationship between vaccinations and increased rates of autism with children. However, researchers studied the relationship of the adjuvants in question (mercury and a preservative called Thimerosal) with rates of autism in children, and found that rates of autism were increasing and rates of Thimesoral use were decreasing. The body of data was not consistent with the hypothesis that increased exposure to Thimerosal-containing vaccines are responsible for the apparent increases in the rates of autism in young children being observed worldwide. Rather, other factors, such as an increased recognition of the disorder in the most and least developmentally delayed children, or genetic factors, were affecting changes (Stehr-Green et al, 2003). Despite research that has disproved common myths about the side effects of vaccinations, there remains a divide that continues to challenge physicians. Indeed, empirical research and data may convince some proponents of anti-immunization, but deep-set philosophies and ingrained beliefs pose a greater challenge.

If research can disprove the false claims about vaccinations, why, then, do religious communities continue to deny vaccinations? In a report analyzing the possible factors for vaccine refusals, negative messages from a third party—specifically, TV, radio, and word of mouth—was the most common reason (Frederickson et. al, 2004). The report data (2004) showed that religious reasons, in fact, were reported less frequently. Therefore, though popular media attention directed toward negative effects of vaccines may play a large role in the perpetuation of misinformation, it may also be an avenue for solutions. A common characteristic shared between vaccine-hesitant patients is the belief that "they can control their child's susceptibility to disease" (Meszaros, 1996)—an idea that may be influenced by negative media coverage. Though negative coverage about vaccines may not go away, the introduction of the benefits of vaccination and a critical analysis of the media's portrayal of vaccinations can provide the public—religious communities, in particular—with a means to make a well-informed decision. Understanding that the concern of religious communities is usually tied to the messages seen on television and different forms of media may be a key starting point to addressing the concerns that perpetuate a "gap" between scientific and religious communities.

Finally, my personal interviews with both pro-vaccine and anti-vaccine friends (for privacy purposes, I will not reveal the names of my sources here) can corroborate the importance of understanding different perspectives before finding solutions to problem. My anti-vaccine, religious friend said that putting foreign chemicals in her body felt like going against the will of God. In further dissecting her argument, it seemed that at the core, she was unsure of vaccine components and safety, and also theologically deviant in her thoughts. In speaking with a pro-vaccine pastor who shared my friends' faith about these beliefs, the pastor asserted that she was theologically wrong about "going against God's will" because their God would want what is best for her, and allowed the creation of vaccines in order to provide for his children. Thus, examining the initial sources of miscommunication—such as theological arguments and personal arguments—can be starting points in which to begin crafting solutions that address the gap. By simply asserting that vaccines are good and creating media that disregards the personal concerns of these often already marginalized communities, no true communication can be accomplished. I argue that the media can play a very large role in lifting barriers between scientific and religious communities, both by presenting the honest concerns of religious communities through documentaries, shorts, etc., and working with scientists to target these communities' *specific* concerns. I argue that finding the starting points in which to attack the issue—such as identifying parental concerns at individual levels during doctor's visits and documenting the trends, or finding the theologically deviant objections that some religious people may be prone to make—can allow for the creation of resources that cater specifically to addressing those concerns. More interdisciplinary research, such as the examination of simple but ingenious advertisement campaigns, may perhaps provide inspiration and options for how to best address the concerns of religious (or simply just anti-vaccine) communities. Focusing on intersections rather than narrow perspectives, in the end, may just yield positive results.

Paolo Joaquin: The Filipino Community and the Achy Breaky Heart

Cardiovascular disease (CVD) is the leading cause of death for all Americans.

This statement already implies that cardiovascular disease affects various communities that reside in the United States. Since CVD is such a deadly disease, many scientist and health groups are trying to address the high death rate by creating methods to living a healthy lifestyle. Healthcare providers and many health officials tell their patients to have good diet and exercise to avoid any symptoms of CVD. However, giving a generic method to all communities within the United States is not effective. The Filipino community in the United States specifically experiences CVD as the leading cause of death. The differing culture of the Filipino community make the solutions of healthcare providers useless. There needs to be a deeper understanding of Filipino culture to make better solutions to the prevention of cardiovascular disease. The community I will focus on is a Filipino community in the Daly City/ San Francisco area. However, it is important to note that the results and experiences of this group of people are specific to this group and are not to be applied to other Filipino communities even if there are similarities.

There are differences among Filipino and American diets. Filipino dishes vary in being very high in sodium, sugar, starch, and fat. However, it is not entirely the fact that Filipino food can be high in those categories. Filipinos will keep eating their traditional food because it is a part of the culture. It is part of the heritage they brought to the United States. "In fact, there is a greater affordability and variety in the local United States markets compared with those in the Philippines." This allows for easier access to foods and larger quantities for a cheaper price. Eating is very important when it comes to family gatherings or parties. One practice that highlights the love of food is something called Pasalubung. "Pasalubung is customary when one person visits another, such as dinner gatherings or after returning from a trip." Even with foods that can contribute to CVD, there are other factors such as stress, a lack of physical activity, and smoking.

Filipino communities within the United States differ from communities who have resided in America for a longer time. There are a lot of factors that make them different from other communities such as diet, language, socio-economic status, educational status, and culture. Before attempting to address the health problems of the Filipino community, it is important to learn about their culture and the reasons behind the prevalence of CVD.

In regards to the larger issue of bridging the gap between science and the public, it is important to not lump the public into one bubble. As I have shown, the Filipino community is a part of the public but is exponentially different because of its cultural practices. Other communities are common in that regard. Many communities within the public differ in their own ways. As a result, it is important for scientist to consider the differences between each community before giving solutions to problems such as cardiovascular disease. Scientist should keep in mind that it is not that simple to create one solution that applies to all. It takes a full understanding of individual communities to create effective and life-saving solutions.

Eliott Ahumada:

From a young age I was taught that in school you had a choice you could either listen to the teacher do all your homework and get a perfect on the test so you could be the smart kid, or listen to the teacher do all your homework and get a low score on your test and be the dumb kid. It wasn't a lesson that any specific person taught me but instead it was the nature of my 1st grade class that instilled this notation on me. Unfortunately for me it only really took a series of tests I was unprepared for that gave me my self defined role I was the "dumb" kid in the class. Its not that I didn't like school or didn't try it's just that I expected that I was going to fail due to the fact that I already had in the past. While over time I learned to get over this mindset it took me a while to come out of the vicious cycle of self doubt and failure caused by the constant pressure of my label that resulted from my standardised educational system. Many of my peers in high school (who came from different elementary schools) that I talked to had similar experiences. This may be due to the fact that us students werent the only ones to give up on us. In his book "Ungifted: Intelligence Redefined" Scott Kaufman cites a study made by Anneta Rattan that states " Teachers who held a fixed theory of intelligence were significantly more likely to diagnose a student as having a low ability based upon a single initially, poor performance. They were also more likely to confront students for their low abilities by saying things like 'it's okay not everyone can be good at math'" When thinking back to when I first started school I remember how highly me and my parents thought of my teacher. While my teacher may have not realized it she was contributing to my self doubt after telling me that I wasn't good at spelling or math. While these little things may not seem big very few of my aforementioned peers ended up pursuing a college degree. Additionally I remember some of my friends who come from essentially the same background as me repeat the same phrase that I too at one point did "I can't do it what do I look like im not good at school"

Andrew Wing:

The issue I want to look at in terms of "Bridging the Gap" between the public and scientists is the kinds of videos and media we are exposing the students of our country to in public schools. These educational videos that students see in school are often seen as a great way to bridge the gap. The videos are witty, funny, and are a great way to teach students science... or so we thought. It turns out that these educational videos aren't quite doing their jobs; students don't learn much more with videos than they do with normal lectures. People speculate that this is because the videos are trying too hard to be fun and interesting, thus their lesson is lost in the translation. Maybe they make things too simple, and students don't get the full picture. However, I can attest that these statements are wrong from my personal experiences in High school. Videos like Bill Nye the Science Guy, Brain Pop, and Nova are actually very educational, they get their point across very well. So if it's not the content that's off, it must be something else, and I think that it's the attitude that both students and teachers have towards these videos. Many teachers, for one reason or another, will show educational videos to their students. Some do it to entertain their students - they see it as a fun way to teach or reinforce a concept to their students; however, most teachers show videos to their students when they're feeling lazy and don't want to teach, or when they're unavailable to come to class. From my experience, when I had a substitute teacher take over a class, the lesson plan was almost always to see a video. Teachers aren't taking these educational videos very seriously, and as a result this attitude is raining down on the students themselves. They see the videos as something shown when teachers have "nothing better to do" or when they are out sick. When students have this sort of attitude when watching videos, they won't give them their full attention and end up not learning much from them. In my experience, my friends and I wouldn't pay attention to these videos and do other things, be it doing work for other classes, looking at our phones, or even taking a nap. This is the roadblock obstructing the bridge between science and the public. In order to make educational videos a real thing that can help bridge the gap, we need to remove this roadblock. What I propose is that we either make the students who are watching the videos do engaging activities, such as worksheets relating to the content (Our PREs in MCB are a great example). Another thing we can try to do on a much larger scale is to try and fix the bad attitude towards videos, and see them as something that can really help us learn more, instead of just being a waste of time for everyone.

CHAPTER 3

PUBLIC ENGAGEMENT IN SCIENCE

A look of fulfillment after finding the connection with a concept that seemed so difficult before.

Chapter 3: Public Engagement of Science

By Sabrina Rentschler

Imagine a scientist. What image comes to mind? Do you imagine someone holding beakers, wearing glasses, working studiously inside a laboratory? Do your thoughts jump to visions of Albert Einstein perhaps? During a study conducted by National Science Teachers Association, 1600 students' grades two to twelve were asked to draw their idea of a scientist as a way to measure the intellectual maturation of preconceived notions regarding scientists, attempting to measure the age in which these associations began to take hold. Only 8% of students portrayed a female scientist even though 60% of the respondents were girls.[1] Another study found similar results; out of 117 fifth graders (57 males and 60 females) 80% perceived scientists as male and 90% white. This chapter will tackle the many ways in which human factors manifest themselves in the scientific community, including how these preconceived notions limit access to minorities in education, research, and innovation.

This chapter also hopes to demonstrate ways in which the procurement of scientific knowledge itself is also wrought with cultural influence. This creates vast variation in understanding among several communities, leading many to wonder: if science is thought to be clear cut and objective, why are there so many differing opinions on newer scientific realisms like vaccinations and global warming? This chapter will attempt to provide an answer to this question through in-depth examinations of the procurement of scientific knowledge in early education, and differences in access to education when considering sociological factors. While it is clear that there appears to be a general stereotypical consensus regarding the racial and gender identity of a scientist, there is also great discrepancy in one's relationship to education and scientific knowledge when considering age, religion, income level, and other cultural factors. Students within this chapter deliver analysis of these issues while also providing their own personal experiences with science through the scope of their own cultural backgrounds.

Students share their own ideas regarding science education and how the scientific profession is formed. They share the ways in which their ideas have evolved over time, eventually leading them to greater understandings that defy widely held stereotypes.

Part 1: "What's in it for me?"

Personal narratives that reflect the benefits the public gain from being engaged with the scientific community.

David Barajas:

The existence of the gap between the public and science creates many difficulties in spreading information relevant to the public's lives and the relay of information from the public to scientists. In relation to vaccines there exists a large amount of controversy and inaccurate information. Parents from multiple backgrounds either strongly support or oppose the use of vaccines for many reasons including religion, personal preference, psychological, fear, etc. I believe that the true reason behind the public's opposition to vaccines lies in the lack of knowledge concerning vaccination. Majority of people do not actually understand what a vaccine is or how it functions in creating antibodies in the immune system. My individual goal is to help spread the information necessary to make educated choices concerning vaccination. The problem in bridging the gap between the public and science lies in the information the public receives and just as importantly how they receive it. People will not find interest in information scientists' release if it does not relate to their lives. We need to show purpose behind the scientific research done, by showing people the importance of science in their daily lives they fill see the significance science holds. Specifically towards vaccination, a large amount of research has been done showing the positive effects of vaccines. Pediatrician Lance Rodewald, M.D., director of the Immunization Services Division of the CDC tells us, "In order for a community to be fully protected against a disease, 80 to 90 percent of its population needs to have been vaccinated." A fear commonly shared by parents who refuse to vaccinate their children is that vaccination can cause autism in children. In 1998 a study was published that first supported this belief, but this study has been proven to be incorrect and was retracted. Multiple studies were done to show how autism was not a response caused by vaccination in children. When the story first came out a widespread of information flourished, even spreading to the Oprah Winfrey Show, but when the studies were released proving this previous study to be incorrect it did not receive the same attention. This lack of attention placed on the truth behind vaccines is now responsible for the lack of knowledge. Parents who still believe this notion keep their children from vaccines because of a fear they gained through false information. This is where the importance lies in change. Those who have the adequate knowledge have a responsibility to share correct information. By pushing towards greater importance placed in the spread of correct scientific information we can save lives, with respect to vaccinations, and help people live a better life as a whole. The process to spread awareness is difficult and takes time, but it is necessary to spread awareness so we as humans can move forward in a positive direction as a whole, bridging the gap between the public and science.

Jaewon (Evelyn) Lee: Science Education and Female Students of Color

I am majoring in environmental biology at UC Berkeley and I immigrated to the United States from South Korea eleven years ago. I have made friends of diverse ethnicity while attending elementary, middle and high school in Los Angeles, California where interesting blend of people of different race, age and gender live together. In terms of academics, my curiosity and capability in science classes motivated me to take physics, biology, botany, chemistry, and organic chemistry courses at UC Berkeley and eventually made me declare my major in biology. However, my female friends who identify themselves as racial minorities (Hispanic and/or African-American background) whom I have kept contact with as early as in fifth grade, often struggled, and did not enjoy taking science classes in middle and high school in comparison to my white and Asian friends. Some of them who I befriended during my freshman year in introductory chemistry classes trailed away from becoming science majors. Why are women of color so disengaged in science education and the scientific community? Are there methods that can be implemented to gravitate and retain of this group, or a subset of "the public," in science?

In the broader theme of public engagement in science, I identify the problem of underrepresentation of women in color in the field of science including physical, life, earth, space and biological sciences. Statistically, the percentage of STEM bachelor's degrees awarded for African American, Native American, and Hispanic women in the United States respectively follows: 5.31%, 0.40%, and 4.33% in comparison to 31.55% for white women (Ong, et al. 174). It is assumed that the percentages are lower for those who completed their degrees solely in science subjects, because the cited figures include the wider scope of STEM. As less female minorities are becoming influential figures in the field, more young women of color are being discouraged from pursuing degrees in the subjects due to the lack of mentors and fellow peers of similar background, culture, and heritage. Fortunately, there are enrichment programs tailored to cultivate more active involvement from women of color in schools throughout the nation. Two of the effective methods include National Science Foundation-sponsored STEM-Enrichment Program (NSF-STEM-EP) and groups on college campuses that serve as supportive communities for Latina/African American/Native American young women who plan to pursue their path to become scientists and instructors in science (Ong, et al. 184). I was fascinated and amazed to read about the existing solutions for the problem of underrepresentation of women of color in science through the paper, (which I cited below) and recently witnessed that clubs like "Society of Women Engineers" and "Hispanic Engineers and Scientists" actively work to increase the number of minority students (especially female students) be engaged in science education at UC Berkeley. Furthermore, I believe that creating a group that connects female science professors and local scientists who are willing to assist fellow female aspiring scientists will be extremely beneficial. College is a place where relationship with people is as important as grades. And in large, public universities like UC Berkeley, it is challenging to find faculty members who are eager to personally help undergraduates. A system where students can earn mentorship and advices through connecting one-to-one with faculty members will be priceless. Another method is to offer affordable science summer camps to promote passion in science topics in girls in middle and high schools in neighborhoods where high percentage of the residents are minorities. The camp would be financially accessible, so no student would encounter barriers to participate. The existing, and possible solution suggested will work as excellent support system for students of similar culture/history and help tremendously to have high retention of minority students in science education. I am certain that the clubs and groups are effective in forming a small community within schools where the students can share comfort and joy with peers of similar goals, interests and background.

In terms of assessment of the effectiveness of the existing solutions mentioned above, Nave, et al. surveyed grade point average of the female minority students who belong in NSF-STEM-EP. They found that the group did better academically in the first-year undergraduate course than their peer male students (Nave, et al. 52). Likewise, an online survey that will ask for anonymous submission of GPA and association in a club/program that is aimed to increase culturally responsible scientists can be distributed to all of the female minority students at UC Berkeley. In this way, it will be possible to analyze the performance of the female minority students majoring in science subjects at UC Berkeley and assess whether the enrichment program that the students belong are successful or not. The survey can be purely asking for the female minority students' subjective comment on behalf of their undergraduate experience. Some of the possible questions can include: If you have any, who are your mentors in college? Where do you find academic support? Do you study with peers of same ethnic background?

Although a continuous work needs to be done to increase female minority students' engagement in science education, the most important factor in continually diversifying the representation of the members will be to promote sustainability. To bridge the gap between the field of science and the women of color, there need to be strong mentoring relationships with faculty that will not only last for the duration of undergraduate years, but also post-graduation and during career work. The mentors will need to continuously have high standards for their female mentees and the relationship will need to build stronger over time to motivate the mentees to achieve more and advance further than their peers. Furthermore, hard work needs to be dedicated in reaching out to younger women of color to recruit and retain them in science. This would be integral in sustaining the diversity represented in the scientific community because it would raise more students that identify themselves as colored females to become scientists of next generation.

Even sometimes, I wonder about the success my hard working non-Asian, non-white friends would have achieved if they had planned to walk on the journey to become scientists. As a UC Berkeley student aspiring to become a biologist, I want to witness more ethnic diversity in the university science departments, pharmaceutical companies, and scientific research institutes. I hope that more schools and organizations in the future will adopt the solutions to the problem of underrepresentation of women of color in the world of science. An effective survey distribution will lead to accurate evaluation of the long-term effectiveness of the solutions. At the end, there will be more dedication and work towards shortening the bridge between the female minority students and the scientific community.

Miriam Juarez: Engaging Low-Income Communities in the Science of Climate Change

Scientific research is not intended to serve the scientific community alone. New insights about disease treatment, viable energy alternatives, and climate change have the potential to benefit communities across socioeconomic status, education levels, culture, and space. However, scientific journals – and in some cases even popular science media – remain extremely inaccessible and unintelligible to groups outside of this elite bubble. It is therefore no surprise that low-income communities are often unaware of the implications of one study or another. As a result, low-income groups are unable to weigh in on seemingly complex issues that are central to their lives. This lack of engagement can have deleterious effects on community interest in science and the ways in which low-income communities perceive science as a discipline.

It is my belief that low-income communities suffer disproportionately from the lack of transparency and accessibility to science literature. They are often labeled as ignorant when in reality they may not have the education or means to comprehend dense science literature. For this reason, the scientific community must adopt strategies to effectively communicate with this subgroup through culturally relevant means. Consider low-income communities in a setting of increasing climate change. By translating the consequences of research findings and reducing calls to action into tangible, short-term benefits, low-income groups will benefit from making beneficial lifestyle changes while simultaneously doing their part to combat pollution and their carbon footprint. Encouraging low-income communities to cut back on their water consumption with the promise of reducing their water bills, for instance, may prove more effective in getting entire households to save water than is listing the species that are in danger of becoming extinct as a result of the drought. Some concerns are more imminent to low-income families than others. Focusing on the values of these communities is key to fostering successful engagement. These sorts of relevant messages can be portrayed in the form of informative videos outlining the positive impacts that lowering energy consumption may have adhering to a family budget.

In this scenario, participation could be measured in the form of tracking water and heating bills in low-income communities throughout an area. Household bills would be compared to expenses documented prior to being informed about the benefits of decreased energy consumption. Previous bills would set a baseline for each household and any increases or decreases would be noted thereafter, where maintaining lower water bills, for instance, would indicate adherence to taking action against climate change. I believe that sustainability will not be an issue, for families will see tangible results after lowering their energy usages and uphold the program themselves.

In conclusion, scientific communities must develop culturally sensitive ways of portraying information to low-income communities for the purpose of stimulating awareness and action in traditionally uninformed populations.

Elizabeth Aguayo: Giving Communities a Voice

The disconnect present between the public and the scientific community is a great impediment toward collaboration, trust, and positive engagement. This disconnect is due in part by a lack of effective communication, especially in regards to the diversity of individuals that encompass what is know as the public. To establish good communication, the scientific community needs to work with each different community in a way that allows them to see the needs of each other and their individual roles in using science not only for themselves but also for the betterment of society.

The public is composed of many different communities, organizations, and types of individuals. Methods of communication and engagement will depend on the community's past exposure to science, education, interests and needs. To put this into perspective I will focus on my own community, a largely Hispanic, low socio-economic community. For an underrepresented community such as mine, the biggest barriers to establishing good communication and connecting the people with the scientific community are a lack of exposure and access to science. To solve this, scientific information needs to be easily accessible to these communities, and presented in such a way that it is easily comprehensible. But this does not mean just giving people information, it means establishing a way in which the community can interact with scientific information, reflect on it, question it, and act upon it.

Communication between the community and scientists thus needs to be based on the idea of mutual learning and benefit. To have this type of communication, Bennet and Jennings (2001) claim, its basis needs to be on "genuine dialogue" with the "aspiration that science and society can work together to shape the future" (p.3). This dialogue is not one based on assumptions but one that would be based on mutual questioning and collaboration. Such dialogue, Sackler (2014) states, needs to give the public "opportunities to ask scientists questions", making them not just "passive receivers of scientific information" (p.100).

To understand how this dialogue can best be established, it is important to get to know the community so that science can be presented with the interests and needs of the community in mind. For this, a "scholarship of integration" among scientists needs to be encouraged, which Boyer (1990) explains, would move knowledge toward engagement by asking questions like "How can [knowledge] be helpful to individuals as well as institutions?" (p.21). Communicating the "how-to", Sackler (2014) claims, can really "empower people" (p.93). Such empowerment is key to connecting my community with the scientist community by demonstrating how science combined with human action can create societal change in their community and in others.

To measure if these methods are effective for the community, a study can be done that measures if there is a significant change in actions that were promoted through scientific engagement. If there were then it could be that the interactions were successful in empowering the people to create change. A before and after survey could also be done in which the community is asked their opinions on science itself and the results can be analyzed to see if there is a more positive opinion of science and its impact in their lives.

It will be eventually be up to the community itself to continue a tradition of scientific engagement. This can be done through parents and schools interacting with their children and students so that younger generations can have the opportunity to start engaging with science. If parents are engaged, a precedent can be set for their children and grandchildren. Lastly, community programs that give access to information and engagement opportunities with members of the scientific community as well as periodic studies on the effectiveness of such engagement will ensure the interests and needs of the community are still being met.

Austin Weinstein: Science in American Political Speech

When John F. Kennedy dared humanity to go to the moon, he was interacting with both science and the public. He took the wonder of space, in its glorious vastness, and connected it with a great unifying sense of public purpose. He was molding the public understanding of science. Public political speech is a place for the public to hear new ideas and be open to changing their mind. When science interacts with this speech, you have a public, rather the audience of the speech, changed, which is crucial to the goal of this course, and this chapter. Understand the interaction between science and speech is vital to the broader understanding of science. This group's charge was to elaborate on "What's in it for me", connecting the experience of the non-scientific public to science itself, giving meaning and clarity to science. That is what President Kennedy was doing when he spoke at Rice University in 1962. Politicians use political speech to address and handle science, and this paper will develop the connections and utility of that oratory in expanding knowledge, and assess it efficacy and sustainability.

When the Soviet Union launched Sputnik in 1957, the world was forever changed. The U.S. felt outmatched, and needed a spark. At Rice University in 1962, President John F. Kennedy provided that spark. He said:
I am delighted that this university is playing a part in putting a man on the moon as part of a great national effort of the United States of America. Many years ago the

great British explorer George Mallory, who was to die on Mount Everest, was asked why did he want to climb it. He said, "Because it is there." Well, space is there, and we're going to climb it, and the moon and the planets are there, and new hopes for knowledge and peace are there. And, therefore, as we set sail we ask God's blessing on the most hazardous and dangerous and greatest adventure on which man has ever embarked. (Kennedy, Sorenson "Rice")

This speech began a new era of American ingenuity and progress. JFK used the wonder of space and the possibility of science to inspire the entire country. The entire country was pushed to care and appreciate the power of science through this oratory. JFK used the wonder of science to get the individual to understand what was truly in it for them. This speech, for the purposes of the project, is Kennedy's intervention. He is intervening in the American political community, his audience, through political speech. By doing this, he is attempting to give an answer to the question "What's in it for me?" to his audience.

Now that the process by which an audience interacts with a political speech is established, parameters need to be established to measure if there has been an impact from that speech. Does Kennedy's answer, America should go to the moon, resonate with the American political public? In this case there are two potential quantifiable metrics that can be used; goals accomplished and public opinion poll changes. The first one, goals, is easily identifiable. One could look at the budgets of NASA, and see that it swelled to its highest percent of the budget subsequent to the speech, in 1964 (Dattaro). One could also look at the direct goal established in the speech "We will go to the moon in this decade" (Kennedy, Sorenson "Rice"). The US did in fact do that, landing on the moon in 1969, one of humanity's greatest achievements. One could say that this oratory did bring action from words. But its effect on the US at large is not measured; an opinion poll would be able to gauge that effect. While Gallup does not published their historical archives, if a poll showed a high percentage of Americans in favor of space exploration in the middle of the 1960s, One could call JFK's Rice speech a quantifiable success. It's important to understand that the using public opinion polls is not totally reliable, because it is impossible to view this political speech in a vacuum. There are barriers to accurate polling in the methodology, such as sample size and pollster bias. Furthermore, there are other factors that would determine the answer to a polling question, like recent unrelated political developments that could skew one's answer to a question like "Do you support America going to the moon?" A voter is less likely to support this if the budget is in sharp deficit, or if they recently lost their job. Thus, using polls to gauge success is only marginally useful.

The sustainability of an intervention is vital to understanding the value of future metrics. Why make another speech about going to the moon if Kennedy's speech didn't make a lasting impact? One potential way to examine sustainability would be to judge the budget of NASA on the long term, adjusted for inflation and budget growth. This would measure the sustainability of Kennedy's intervention on the long term, but it would also be a metric for all the other political interventions into science and theirs impacts, such as the Challenger disaster. There was an immediate spike after Kennedy's initial push, but over time, budget woes began to arise, and funding petered off.

A retrospective analysis of one speech can reveal a great deal about how the political public views science. In speaking about going to the moon, Kennedy intervened in the American voting community, asking them to consider an answer to "What's in it for them?" in regards to science. The impacts of this intervention can encourage or discourage certain policy decisions, and affect America scientific policy—in addition to inspiring communities all over. A president reached into the American zeitgeist and turned it towards the moon, and towards science.

Liang-Han Tai (Chase): Science as Performance: Problems and Solutions

Being one of the most influential sociologists in the twentieth century, Ervin Goffman once said that a "performance" may be broadly defined as any activity if given participants on a given occasion. Decades after Goffman's commentary, as the mass media continuously develop and diversify, topics regarding scientific research or popular science have become more accessible. Meanwhile, the way in which science is presented could be observed and examined as a form of performance, while the audience remains skeptical when science-related information is provided through the media.

The audience, referred to as "the public" in this case, is strictly defined as the individuals with access to scientific content in the media. If science were treated as performance, the primary objective ought to be effectively communicating with its audience. Therefore, the "setting" becomes very important in terms of framing the issues. A theater production is more convincing when its theatrical setting is realistic and engageable. Today, social media often have the ability to convince their audience because they have created a forum filled with ideas and perspectives. Such liberty of expressing opinions has made the platforms seem more knowledgeable, yet at the same time, it has decreased both the credibility and responsibility of media.

Other than the setting itself, the role of actors is also prominent as it ranges from scientists, reporters, writers, to viewers and other spectators and observers. With such a high demographic variance, the scientific content could be easily distorted and misfocused, which has a close association with the size of the gap between producers and receivers. From the theatrical standpoint, the most crucial and advantageous practice would be to encourage and promote an alternative common ground where both sides can attain a certain level of consensus with lower barriers to entry, and further negotiate and interact.

As we identify the problems, it is necessary to examine the sources. One of the main reasons result from the mindset of categorizing and over-simplifying science as a school of thought rather than a constantly changing phenomenon. The lack of exposure could also be a an issue leading to the problems because the fields of research and development are often limited to the elites, which could potentially decrease not only the audience's interest in science, but also the science community's effort in creating active engagement and collaboration with its audience. These problems, however, are indeed resolvable using certain methods of assessment following intervention.

The above methods are referred to as the "instruments" of theatrical production, with the purpose of enhancing the quality and efficiency of the performance. Some of the most common approaches include the use of media, platforms outside of K-12 curriculum, and highlighting personal benefits through various surveys. As previously mentioned, media, especially social media, have become one of the prominent resources for public engagement from politicians, entertainers, and education facilitators. Therefore, even simply increasing the amount of science-related news and articles has the capability of fostering a beneficial atmosphere as if forces the audience to receive the information at the first place. Yet, the information needs to be not only accurate but open to brainstorming, comments, and sometimes criticisms. Needless to say, other platforms such as after-school courses, summer camps, and conferences regarding current trending topics in the field(s) of science can be possible outcomes solving the problem of lacking in access and involvement. Lastly, the driving force behind any human behavior is the motivation towards personal interest. In order to incentivize its audience to acknowledge and further support the scientists, as well as the science community as a whole, one may underscore the benefits and advantages of public engagement. For instance, it could be difficult to convince a housekeeper to help promote mechanical engineering fundings due to his/her lack of interest. However, if this housekeeper can relate him/herself to the potential new products, and realized the benefits resulted from these mechanical engineering discoveries, such as a less time-consuming vacuum cleaner, dish washer, or any domestic appliances, the science communities are likely to receive more support from its audience. More importantly, these communities would be able to attract different audiences rather than specific, targeted groups among their information recipients.

Even with initial doubts and reservations, most of the recipients, or viewers, are easily convinced without solid proofs or validation, which signifies the severe problem of blind obedience, as well as following without recognizing. Nevertheless, by looking at science as an act, or a composition of various acts, there are many aspects to be dissected under theatrical lenses and drawn attention to, in an environment where facts are twisted and limited. Furthermore, the public could cease the debate of how issues emerge, and investigate on how the components within the process could be changed, facilitated, and later prioritized.

To further sustain the outcome, more questions must be examined. The most important among all: How is the public treating the information provided? What is the public's first, second, and final reaction toward the content change throughout a given period of time? And how "much", if measurable, does the public trust the information provided. Ultimately, treating science as a performance, it is rather less challenging and more comprehensible to convey what the public *can do* for the public, and essentially entails what is "in us."

Steven Yu:

From my experiences of being in both the public and scientific community, I can understand why there is such a gap between the two. I refer to the public here as the group of people who are not knowledgeable in a certain area and the experts who will be referred to as the scientific community. In both communities there are differences in beliefs and understanding. There's even a spectrum of perspectives within each specific community so in order for one to target the "public", it is difficult (Matsui). From my experiences with the public which includes some of my friends, they tend to pit religion and science against each other despite the fact that they are both different in many aspects. They say things like "I believe in science" as if it were a religion, but science is a process in which people use evidence to support claims (Padian). Even on the Internet, the public continues to pit science against religion without anybody really trying to point out that they are two completely different things. The idea that science is more reliable than religion helps project the image that science should always be trusted because it is quantitative, rigorous, and objective unlike the humanities, which use mostly qualitative data to describe phenomena. The public's ignorance to the process of science further limits their access to actual research because the public depends on social media or other forms of communication to help them "believe" in the science that they try to preach. Even then, the science that they are exposed to is subjective because someone else who wrote the article based on their interpretation (Matsui). The result is that people end up taking it for face value. When I was doing the social media project and I sent an article out to my friends to read it, many of them said it was amazing or that it was cool. They probably did not know that the article was based on an actual paper from a journal. Due to that knowledge limit, their interest stops there because they have no other way to verify the information provided to them. This is why the first step to educating the public about science is to have them understand the process behind how science is done so that they can be skeptical about the work rather than taking everything in. An example would be how the media influenced the belief that vaccines were harmful to children and would cause autism, but that paper has since been taken back. However, people still continue to believe that the science was definite and that they should believe it, but science is often overturned by new ideas which they don't realize. This is where the scientific community comes in and performs outreach to educate those who still believe in old news. Outreach could involve community lectures or scientists and the public can gather at a museum to have open dialogue about the things that are going on in the field. This way the public would learn more about where their taxes are going towards and how it affects them personally. The public can then use this information to make informed decisions that benefits themselves and their community. This open dialogue could also allow scientists to focus their research on things that the public cares about and would benefit society as a whole. With science and the public interacting, the process of improving the quality of life or even solve large overarching problems would be easier. In the long run, this kind of interaction would benefit both communities, but only as long as people are able to see the long term benefits of being interconnected with each other.

Josephine Espinoza:

Social, political, and economic determinants make substantial contributions to an individual's health. Low socioeconomic status (SES), for example, often measured as an individual's combined educational level, income, and occupation, is a particularly important predictor of an individual's life span development, physical health, and psychological well-being. There are various reasons for this: low SES individuals often have greater difficulty accessing and purchasing healthy food, are more at risk of exposure to damaging agents in the environment, are less likely to have access to routine health care, and tend to live in neighborhoods with higher prevalence rates of violence. These factors translate into these individuals having a higher likelihood of poor health behaviors (such as, poor diet, limited exercise, greater alcohol or drug consumption and reliance, smoking, etc.) and consequently, poor health. It is an unfortunate reality that many of these outcomes can also be applied to other communities, such as, people of color and the LGBTQ community. We know, thus, those who are at the greatest risk for largely preventable diseases (e.g. lung cancer, cardiovascular disease, etc.) in the United States are also at the greatest social, political, and economic disadvantage. For this reason, among others, we need more of an emphasis on preventive care and services in our health care system to treat common health problems and promote healthy lifestyle behaviors, such as adequate exercise, healthy food, reduced alcohol, tobacco, and drug consumption, better psychological well-being, healthier relationships and social support systems, decreased stress, proper immunization, more screenings etc. Primary care, the area of our health care system we generally attribute preventive care to, however, is continually shrinking as a workforce. Primary care providers are often overworked and undercompensated in comparison to others in specialized fields making it increasingly less common for doctors to pursue this area of care. Further, due to time constraints, primary care doctors are not as able to foster a bidirectional dialogue with their patients, often taking on more of a paternalistic rather than collaborative role. This is not always the most effective means of communication among two individuals.

In the face of all these issues and in an attempt to emphasize and improve preventive care, we need to ensure there is better understanding of the importance of health behaviors on our health in the United States, especially among communities of which are disproportionately affected by preventable diseases. This is where an individual's engagement is of particular relevance—for the improvement of one's own health and one's communities' health. In engaging, whether it be through greater, more effective dialogue with one's doctor, the pursuit of health behavior knowledge available, etc., an individual can better be able to understand the complexity of diseases, their onset, and progression. They may learn, for example, that having a genetic predisposition to, say, diabetes can be alarming, sure, but need not be the end-all, be-all. They may learn that in making positive health behavior adjustments to their lifestyle, they may greatly reduce their risk altogether. We need to improve how we communicate risk and should stress self-efficacy among these individuals. A fatalistic outlook on disease risk and progression does little to improve patient motivation and adherence to healthier lifestyle changes. Instead, we should emphasize self-efficacy to these individuals—communicate to them both that healthy lifestyle changes are beneficial and that they have the capacity to carry out these changes.

As mentioned previously, there are determinants, often out of the control of these communities, of which affect their health, and engagement in science for the advocacy of oneself and one's community can only go so far. Though we should continue to advocate for short-term, immediate goals for these communities, we should also strive to seek solutions for the long-term barriers – social, political, and economic – they face as it is these influences on health behaviors of which maintain and reinforce the gap. We need to oversee and implement government and business policy in order to effectively combat the gap and sustain the short-term change engagement creates. For example, we need more compensation for primary care doctors in health care, better policies around environmental hazards, more funding for recreational spaces in underserved areas where individuals could exercise, etc. In doing so, it is our hope we could improve the health of these individuals and of this nation as a whole.

Part 2: "How can we engage?"

Our objective is to explore different methods of outreach to lure the general public into a positive and voluntary relationship with science. We identify social, cultural, and educational factors that establish communication and relationships between the public and science.

Nick Garelis:

The majority of the population in Western society tends to have an aversion to STEM -- a natural reaction to the way "science" is typically presented in school; disaffected youths soon become antipathetic adults, and the potential for successful engagement is squandered. Two-thirds of the students in US high schools report being bored during class,[i] and finding ways to break through that boredom is essential; it's no coincidence that US scores in STEM lag far behind those of many other countries.[ii] Getting students interested in STEM at an early age would help overcome that aversion; without tapping into that available pool of young minds, we are wasting a valuable resource.

Many inner-city and low-income students don't have access to the popular "Science Camps" that turn STEM concepts into an exciting, accessible adventure; the only interface they have with "science" is being force-fed the Periodic Table or made to memorize incomprehensible formulas – abstract information that bears no tangible relation to the world they inhabit and the problems they face. One way to capture their attention is through hands-on projects that solve real-life everyday problems: how do you jump start your car if the battery dies? How do you put out a grease fire? If lost in the wilderness or -- on a larger scale -- in the event of a natural disaster, how will you find potable drinking water and construct a shelter? By solving these problems firsthand, students absorb key STEM elements while learning to harness electrical current, anticipate and control the interactions of chemicals, apply the properties of condensation and filtration, and design the system that enables these processes. This approach gives students hands-on experience they can't get from a lecture or textbook, making STEM relevant to their needs.

One high school in Indiana experimented with this idea, encouraging students to design and construct a portable disaster-relief shelter with a water purification system and a renewable power source; their invention was then presented to President Obama at the White House Science Fair. In California, the Oakland Unified School District is enlarging its high school STEM curriculum and working to establish STEM centers in its middle schools to help students engage in applied learning and thereby gain fundamental experience in such fields as engineering design.[iii] By being able to identify real-life problems, and by testing possible solutions to those problems, students can develop vital skills in problem-solving, as well as in teamwork, communication, and leadership, which will help prepare them for college and for careers in the STEM-dependent future. The importance of creating an integrated STEM curriculum that encourages scientific inquiry and applied learning has been recognized, and many organizations are stepping up to help. One top provider of STEM curricular programs in schools, "Project Lead the Way," announced a $6 million partnership with Chevron to help fund these high school and middle school programs.[iv]

The world's needs have evolved, and students' skills and education must also evolve. The first step is developing a user-friendly relationship with STEM; after that, the rest will follow. For the next generation of students to embrace and succeed in STEM, they need trial-and-error experience in creating, constructing and refining new solutions and inventions. Demystifying STEM by giving students an early-on working relationship with basic skills will provide them with a broad base of experience from which to draw, whether they choose to pursue careers in STEM or not; this will make them more receptive to engaging with other sci-tech issues they face in the course their everyday lives, such as understanding the importance of vaccinations or addressing climate change -- genuine problems and decisions that directly affect their futures.

Exploring activities that address real-world problems can inspire students to willingly engage with and participate in STEM; with this approach, students recognize the disciplines as valuable and accessible tools they can use to chart and navigate their lives, rather than as a boring, confusing, and cumbersome burden for which they feel an aversion.

Deena Abdelhalim:

In order to engage people in science, whether it is medical or research, our first step needs to be in educating. If people are not educated in at least basic scientific knowledge, they will not be interested to continue pursuing interests in the subject.

Education is not restricted according to age and thus people are not exempt from being knowledgeable about basic and current topics. In fact, this is what is driving the problem between the public and science because some believe that there is no reason to be educated when their opinions will not be heard because power is placed with the scientists. However, the reason behind being educated is so that people can independently decipher content accurately without being susceptible to false information. My social media project was a prime example of "blindly" accepting the facts where those in my experiment believed a study on the effects of marijuana on the brain, when in fact, the study was poorly executed and the results were highly flawed. My specific part in the chapter about "How can we engage" is to see how people naturally accept articles that coincide with their beliefs as opposed to interpreting the validity of the research and relating it to the necessity of education as groundwork and a basic platform before engagement in science can be achieved.

  This issue of people not being well educated on important scientific matters is not a problem that can be immediately solved. To try and address the issue of improper education, we would have to implement change at the beginning levels of institution, where children are first introduced to scientific material. This is a critical time because many students lose interest in subjects from an early stage in their lives but if the material is taught in an engaging and interactive way, the children might be more inclined to pursue those subjects. But at times, the issue is not that people are not educated at all, but rather have a biased opinion regarding topics, which skews their perspective. This is typically seen in the older generations which is why altering their mindset would be more difficult than implementing change in the younger generations from the beginning of their education. Through my experience, I have found that people tend to only read articles that coincide with their interests. This however also translates to what they view as right. In my experiment with the topic of marijuana, if I were to provide an article about legalization to those who agree with the legalization of marijuana versus those who disagree, their responses to the article would have been vastly different. This is because, psychologically, we agree and enjoy articles that support our perspective. We can try to reduce this problem at an institutional level by encouraging more debates that require taking the opposing side of one's beliefs; this could hopefully lead to more engaging discussions on a broader scale with friends and families. While debates are sometimes applied in classes, in my opinion, they are intended to encourage class participation as opposed to stress the importance of understanding both sides of an issue and qualitatively being able to reason through both. It is integral to learn to accept and understand various sides of an issue because that provides a foundation of knowledge as opposed to opinion. While this is difficult to gauge for success, the more discussion and conversation that is stimulated between people would provide them with more information that is not limited to their own opinions. Tracking the success of improving the groundwork of education in the younger generation can be monitored over several years to see if more children continued in science fields after being introduced to the subject in more interactive and engaging ways. Education, whether it is introducing more engaging teaching methods or understanding other views, is the necessary foundation needed before we are able to achieve engagement in science.

Charles Li: It's A Two-Way Street: Beyond the Public Understanding of Science

The social media projects of both my peers and my own have allowed me to realize how difficult it is to induce engagement in science even through what we may perceive as an accessible means of mass communication. Though social media is an effective way of disseminating ideas, sharing information, and voicing opinions in a timely and interactive manner, it appears that an individual's response to scientific information is often governed not by the technological medium through which science is conveyed but rather determined by personal biases and internalized conceptualizations of scientific issues. I am interested in not only exploring why social media is not as effective as we had hypothesized it to be but also what the underlying factors that inhibit science communication are. We should begin by elucidating the goal of science communication: is bridging "the gap" about fostering trust between scientists and the public? If so, how is trust and mutual respect ever reconciled with the healthy skepticism? Furthermore, if we assume that a successful "bridge" between scientists and nonscientists is one that allows the public to appreciate science as a valuable tool of inquiry with significant social benefits while encouraging scientists to recognize the importance of not alienating those who do not understand their work, is there a singularly effective medium through which we can bridge the gap? Or is there no quick fix to a problem that ultimately lies, as the rest of my group has argued, with more intrinsic factors like one's level of formal education, value systems, and culture? Because it is difficult to provide concrete answers to such broad questions, I will focus on the current controversies surrounding the consumption of genetically modified organisms (GMOs) as a model paradigm through which the roles of trust, skepticism, personal bias, and scientific truth in the public understanding of science can be explored. I will be analyzing the prose of Mark Lynas' recent *New York Times* editorial "How I Got Converted to G.M.O. Food" to better understand the factors that influence the popular perceptions of scientific issues.

Many of the barriers and bridges surrounding any instance of communication are fundamentally shaped by the intentions of the communicator himself. Lynas, an outspoken environmental activist and researcher at the Cornell Alliance for Science, seems to be motivated by the noble objective of urging rational discourse on the issue of GMO food safety because there is, as Lynas believes, little empirical evidence for health detriments directly caused by GMOs. He points to the absurdity of fearing crops that have been proven to make farming efficient through increased resistance to extreme environmental conditions and pests and claims that any opposition to the use of GMOs is supported only by emotional speculation.

Ironically, in urging ration, Lynas overlooks ration*ale*. Like many others who attempt to communicate science, he insists that readers "look at the science", "look at the facts", or "look at the evidence" without providing an appropriate background for understanding what the biology behind the safety of GMOs is to begin with or establishing a framework that enables readers to evaluate scientific data relevant to this topic. Simply stating that there is a high level of "scientific consensus" is not only an incredibly ineffective way to facilitate a rational debate on a controversial issue but also inappropriately assigns an inherent superiority to the notion of "science" that often leads to major barriers in communication. It can be argued that it is difficult, if not impossible, to impart an appropriate paradigm for interpreting scientific findings in a one-page editorial because such a fluency in scientific literacy would take years of formal education to acquire and develop. However, there are more immediate problems that create barriers in science communication regardless of one's level of technical knowledge.

In an attempt to bring his readers closer to "science", Lynas establishes a binary between "scientists" and the "public" that disproportionately assigns automatic intellectual value and social merit to conventional scientists and the technical knowledge they produce. Drawing an analogy between the seemingly undeniable phenomenon that is the rapid occurrence of global warming and the safety of GMOs, Lynas states, "After writing two books on the science of climate change, I decided I could no longer continue taking a pro-science position on global warming and an anti-science position on G.M.O.s. There is an equivalent level of scientific consensus on both issues, I realized, that climate change is real and genetically modified foods are safe. I could not defend the expert consensus on one issue while opposing it on the other." It is remarkable to see how that the author uses the terms "pro-science" and "anti-science" as if science was a stagnant, discrete, and singular *moral* subject instead of a systematic way of inquiry that is welcome to skepticism and appropriate changes to its repertoire of hypotheses and theories. Interestingly enough, Lynas then mentions that the outright rejection of GMOs is an act of "undermining public understanding of science", citing a Pew Research Center study that "showed a greater gap between scientists and the public on G.M.O.s than on any other scientific controversy." The author again mentions the popular perception of science as if there was only one specific population interpreting one precise, easily accessible, and irrefutable body of knowledge known as "science." Such an attitude is counterproductive because it not only fails to acknowledge the many ways in which scientific data can be interpreted but also ignores the inherently *un*known qualities surrounding almost every scientific phenomenon in existence.

These communicative barriers, however, can very well be removed precisely because of a "public" that is able to independently supply the needed skepticisms to those who identify themselves as messengers between the heavenly world of science and the earthly existence of non-scientists. An example of such critical thought is found a well-articulated comment written on the NYTimes.com discussion board by a user named "Somewhere":

"I appreciate the writer's analogy to climate change and his appeal to an educated reader's desire to consider herself thoughtful, rational, and reasonable. However, making changes to the genetic ecosystem, no matter how well studied, well considered, and effective, is the very essence of the slippery slope. Taking measures to curb emissions is not engaging with a system of life - defined by self replication, mutation, evolution. Consider almost any scientific pursuit. In experimentation, side effects and unexpected outcomes are studied, precisely because they are common occurrences. It may seem, and may be, short sighted and reactionary to stand in the way of a pest resistant purple vegetable...

For the same reason that it is possible to predict the outcomes of climate engineering, but still acceptable to question science's ability to fully understand all that may go amiss, I wonder if opposition to GMO in the food chain isn't wisely conservative, and protective of an essentially unpredictable process: life."

In our efforts to facilitate science communication, we have focused on ways in which non-scientists can better appreciate the advanced technical work of scientists. In doing so, however, it is just as important to realize that lacking a high level of formal scientific education does not translate to being deficient in critical cognition or analytical thought. Perhaps it is time to examine science's understanding of "the public" in order to allow for a productive discourse that appropriately acknowledges the complex, nuanced, and thoughtful beings that we all are.

Krystin Ventura: Young Women and Science During the Critical Period

Two facts about our society are becoming increasingly difficult to ignore: American education is falling behind compared to those of other developed nations around the world, and American women are severely underrepresented in most scientific communities. The former is so significant that when we as a group tried to address the problem of less-than-ideal "public engagement in science," we kept tracing our proposed solutions back to the education system. Notably, most topics we wanted to write on individually overlapped at American education. Generally, our thoughts on how we could increase engagement went something like this:

Tackling the education system as a whole would have been considerably more difficult. As stated above, there are so many factors that have influenced the drop in America's competitive edge in the global educational scale: socioeconomic factors, political changes, cultural shifts, technological advances, globalization, and so on. Instead, we focused our ideas and aimed at one particularly vulnerable category of Americans: female students in early teenage years. This appears to be the "critical point" in the lives of many girls, where social pressures and cultural norms often affect their decisions to pursue education and careers in the STEM (science, technology, engineering, mathematics) regions. Our culture perpetuates this by impressing upon children that intelligence and "nerdiness" are character flaws. This is especially true for girls, who are encouraged to avoid anything deemed masculine (as science often is) that would compromise their perceived femininity to others. Pushing away half of our population from these subjects, in whatever way, undoubtedly affects scientific engagement from society. How can Americans maintain a scientific stronghold if we almost actively discourage half of our students from increasingly more competitive fields?

This issue is especially personal for me. As both a woman and an ethnic minority, I feel that I was discouraged from pursuing an education and career in science much more than my male peers. In middle school, a male teacher blurted that science is often too rigorous for most women, and that there's a reason it's a male-dominated area. When I got to college, a counselor told me that "girls like me" generally don't end up with hard technical degrees, and that maybe I should rethink my academic decisions. As someone who recognizes the sexism I will undoubtedly continue to encounter in my education and beyond, I'm still fearful that it will take a toll on my life and my own confidence. If someone like me, who has already made the decision to stick with my dreams of becoming a bioengineer with a Berkeley degree, is fearful, imagine what this could do to a similar young girl who wants to follow her own dreams. With this in mind, it isn't difficult to think of the many ways a woman would be discouraged from STEM fields. That's why I want to help foster a far-reaching environment in which women are no longer pushed to shy away from anything they want to do. I think creating this encouraging environment would affect more than just our education systems, but would increase confidence and tenacity of all American youth.

We also wanted to show the relevance of science in our daily lives. One of the issues that contributes to the lack of interest in the subject is that our education system ignores hands-on learning in favor of textbooks, standardized tests, and memorization of formulas. Instead of just these bland approaches of learning, we wanted to show how interesting it can be to discover the ways in which our muscles produce kinetic energy when we run and the chemistry of cake ingredients change when we throw them in the oven. This doesn't just benefit girls, and encourages a new way of approaching scientific learning all around. The effects of our ideas are in no way immediate. This was one of the most difficult problems to address, and we as a society need to realize that any changes we make in our perception of science will only be observable after decades of persistent efforts to change how people at all ages view science.

As a group, though, we remain hopeful that current trends can be reversed. We as a society have the capability of producing more intelligent and enthusiastic students of life, at any age. Science, as broad of a term as it is, is something that should always remain a topic of interest for anyone at any age. Our knowledge of it doesn't become irrelevant once we end our time in life as an academic student and become part of the workforce. We remain voters, decision makers, and role models—wouldn't keeping an active fascination in discovery and learning make us better at all these activities?

Lorene Cudjoe: *A Focus on Middle School Girls*: Cultural Humility as the Key to Destigmatize Science

To engage is the act of attracting or involving someone or something. How then can we, as scientists, attract and involve the public in our research, our findings, and our labs? I believe it must begin with a paradigm shift. Scientist must first develop cultural humility, which allows them to place equal significance on the interests of the public and that of their own. By also developing cultural humility, the script is also flipped into the scientific understanding of the public whereby there is an initiation of a reciprocal relationship or a bridge built between the gap of both communities.

How might cultural humility look like in practice? First, it is important to define 'public'—which may vary from scientist to scientist—in order to know who one is attempting to involve. For example, we defined our public as middle school girls. Secondly, it is important to consider the biases that may currently inform science in order to keep science as objective as possible. Lastly, one must consider the pressing issues of the group one defined as the public in order to collaboratively and collectively weave together the interests of both the scientist and that of the public.

How do we know that incorporating cultural humility in science is an effective way of engaging the middle school girls? Through our video, we have been able to address factors that may play into why women account for a small percentage of those in STEM careers. Some being the masculinization of science, the incorrect traditional view that science is wearing lab coats and mixing chemicals in test tubes, and the lack of engaging learning styles in the classroom. By doing so, we have addressed some of the pressing issues in our society as it pertains to women in science and believe that should all our efforts be implemented, then statistics will show an increase in women in STEM related careers, as it has in recent years as society is slowly becoming more progressive.

Jonathan Homidan: *Engage:* Media Intervention and Sociology

The core theme of my group's assignment is defined as to suggest meaningful and potential routes that facilitators can take to ensure that the knowledge 'gap' between the public and science-related academia will shrink. However, before we can address and analyze which tools we can use to engage with the majority of the public, we must first ask why the status quo is failing. Coming from my own experience—and that of many of my peers and group members—there existed a time where science ran rampant on every child's mind; dreams of space, animals, and wild concoctions were frequent and often shared with others. However, there became a point where science separated from social dogma, and, thus, became slowly rejected; these dreams and vivid imaginations filled with science became few and far between. Somewhere along the line, the glamor of intelligence and discovery began to fade; the spirit of the science failed us. Though many of the youth, headstrong, gave no heed to the thoughts of their peers and continued along their path, many succumbed to peer pressures and began devoting their time to other subjects. Years and years of this constant trend led to the gradual creation of this social dogma—that women were not suited for science, and that almost every career path and choice would lead them to green pastures.

Though, of course, it is a bit of a generalization to assert that dreams of careers in science were largely deleted because of sociology, it seems to be the case. Young students, especially females, became jaded of science because the field lay so far away from the fields of glory and recognition that humans, at all ages, naturally crave. Instead, they dream now to be world-famous athletes, singers, and movie stars, among other fields also glorified by the media. The media, an extremely powerful tool, fails at its job of making scientists role models for the younger generations. However, minor media intervention is surely not enough; the problem continues with the fact that from a very young age, science is closely related to schoolwork, while sports and entertainment are attributed to fun and leisure.

Add that to the fact that a career in science requires immense discipline and brilliance (because of the many years it takes to attain relevancy in the community as well as the increasing level of difficulty with each step), only reserved for the elite of minds and attitudes alike, we are presented with a large problem. It seems that the only way to actually engage with the public is to target and change the mindsets of the younger generations. Parents, only seeking to create better futures, opportunities, and lives for their progeny, will be swept along with the tide. The only way to truly significantly engage with the younger generation is to use an often-evil, corrupt, and misguided tool for our benefit. When media is able to appreciate science and highlight the importance of those in the scientific community and portray to younger audiences that scientists are who everyone should aspire to be, that is when change is born.

Thus, the issue becomes finding an attainable solution to combat the ongoing societal perception of science and then maintaining the solution in order to change this perception. I suggest that we use the media and extend its coverage of the science community so that the younger generation can see that scientists, too, can be superheroes and role models alike. Furthermore, constant reminders that science is not just a subject, but also a relevant art existent in nearly every single thing and activity we can see and perform will be just as necessary. Society must see that scientists are more than madmen slaving away to create pharmaceuticals that are expensive and may someday prove useful; they are heroes of their own right, helping us understand and unravel the old, tangled web called life on our planet Earth.

Sarah Cho: Interactive Methods, Mandatory Education

For many years, the scientists executed numerous attempts to engage the public in science, but there were many incidents of failure. In the articles, "Parents Who Shun Vaccines Tend to Cluster, Boosting Children's Risk" and "What's Up With Parents Who Don't Vaccinate Their Children", science communities sought to engage the public (*parents who have children*) in decision-making concerning vaccinations. These articles convey motivations of not only scientists who seek this public's interest, but also the public's motivation of engaging in science.

In the United States, the science community attempted to persuade parents to vaccinate their children because "vaccines are among the safest, most effective ways to protect children from major communicable diseases"; however, "parents still doubt this" and refuse to vaccinate their children (Gross). Unexpectedly, this open space to freely choose increased the engagement in science, because people began to research the pros and cons of vaccinating their children; but still, many parents preferred not to vaccinate their children. However, in Nigeria, "rumors spread that polio vaccines were surreptitious sterilization efforts. That led to a boycott of the vaccine in 2003 and a resurgence in the poliovirus three years later" (Brink). In the contrary, this clearly exemplifies the lack of engagement in science. These situations causes the readers to question under what circumstances is it possible to not only engage the public in scientific issues, but also create constructive ways in help decision-making. Because through these incidents, readers are able to understand that a greater engagement in science through open decision-making does not necessarily mean that the public will make the right choice.

During my first high school registration, the school had vaccination requirements. If these requirements were not fulfilled, students could not attend the school. Therefore, after finding a clinic, I took the vaccinations that I needed and registered for the high school. This was not a problem for me but, a big issue for a friend that I went to school together since kindergarten. Shockingly, I did not see her in high school. She told me that her parents did not allow her to take the vaccinations because of the tragic stories that they have heard about vaccines. There are many difficulties in engaging parents with science positively. Especially, when parents are willing to strongly believe the rumors rather than investigating the facts—the science. The outcome could have been different if the school had educated the parents or at least given them important reasons why their children needed to take these vaccinations. When I was notified to take these vaccinations, I just received a paper with the list of vaccinations that was required. There were no instructions of where to take the vaccinations, or why we were required to take them. When parents were given the choice to choose, it would have been helpful if the school *guided* the parents toward the correct objective decision rather than being swayed by misconceptions.

These parent's attitude toward vaccinations were similar to the suspicion and mistrust of polio inoculations in Nigeria. This demonstrates the life-or-death consequences of decisions made with insufficient knowledge. Scientists need to increase the overall awareness and involvement of science, in the hopes of preventing unnecessary tragedies exactly like this—"resurgence in the poliovirus three years later" (Brink). The Nigeria incident points out that not engaging with science can have very serious consequences, and that is exactly why efforts must be made to educate and involve the public (parents or anyone taking vaccinations). The vaccine debate is an important example of why public engagement in science is even more so important.

It illustrates just how disconnected many individuals are from the scientific community. Even broader, it demonstrates the mistrust that many have in scientific research and institutions. The polio vaccine in Nigeria gives a prime example of the consequences when people are so disconnected from the facts.

We need to try to alleviate the disparity between science and the parents making such imperative decision for themselves and their children. Scientists need to come up with methods to get more parents to trust and become more interested in scientific research in order to increase scientific engagement in the public. There needs to be methods that can make science more inviting and accepting to more of the parents with children who need to take vaccinations. Possibly one means is by educating the public in an easy and understandable matter. Such as creating simple interactive pamphlets for people who do not have access to the internet, and an interactive website that both students and parents can utilize. It would be important that the pamphlet and website be easily accessible, interactive, and also mandatory for both students and parents in order to send their children to school. In the website/pamphlet, it would explain what vaccines are, what they can do, and why people need it in a friendly and fun manner for all ages. It would be interactive with games about vaccinations for young children, and also include research papers and questions for parents. I personally did not understand how shots worked until middle school. Therefore, it would also be helpful to include clarifications of all misconceptions that have come up about vaccinations and why it works. In this pamphlet/website, it would also be important to include where people can get vaccination shots in their area. This would be sustainable and effective if vaccinations were required to be educated in school at a young age through the use of this website or pamphlet. It could be effective because by educating students at a young age, they will be well-informed about vaccinations by the time they become parents.

CHAPTER 4

Scientists are people, too

To maintain a balance the impact between scientists and the public is a current that moves both ways.

Chapter 4: Scientists are people, too: Critical Review of Science

By Danny Lee

Edited by Julie Mendoza

Individuals who make up the "scientific community," or what it is understood as, do not solely exist in a bubble; they can be found at PTA meetings, shopping at the Farmer's Market, or riding bikes through bustling urban cities. Though this may seem obvious, remembering the identities of working men and women inside *and* outside of the laboratory is essential to fully understanding their work within science.

Observing how individuals interact within society while outside of their field reminds us of the interdisciplinary space that experiments and research are woven into. Each field relies on one another to succeed and perform effectively. The relationship between science and business, science and media, media and education, and so on requires a multicultural lens and the distinctly human ability to reach beyond a single discipline. Thus, science as a collective cannot exist within the confines of a bubble.

Experimentation can be described as the act of discovering. Its purpose is not to define itself by coming to a concise conclusion. It is instead given the responsibility of materializing reliable data. Rather it is the noble role of the remaining disciplines to decide on an effective approach in delivering this information to the public. Writers, publishers, media outlets, business conglomerates, teachers, and whoever has the authority to voice the purpose of laboratories unknowingly impact the reputations of working individuals within these laboratories, otherwise known as scientists. This explains the importance of peeling back the layers of an institution in order to properly understand its functions as an entity. It is tangled with many other factors that stem from its relationships with other disciplines. Seeing each layer separately may reveal the true status of an institution and ways it can be improved. In this chapter, we will discuss the concerns, conditions, and solutions regarding what and who the scientific community is made out of and its context in society.

Scientists are People, Too!

Taking Off The Lab Coat

🔊 WHOSE VOICES ARE HEARD

Underrepresented Minorities in STEM, Experimental Biases, Stereotypes in the Scientific Community, Resources to Combat Discrimination in the STEM Workplace and Classroom, Lack of Universal Vocabulary as a Barrier to Scientific Literacy, Commodification of Scientific Discovery

👤 PEOPLE IN THE INDUSTRY

Unsung Heroes in Science, Decisions of Researchers and Doctors within the field and the Implications of the Hippocratic Oath, Discussion of Harm Reduction for Drug Users, Exploration of the Public Health Profession

♡ ETHICAL CONCERNS

Insight into ethical questions concerning technology, Prenatal Engineering, Artificial Insemination, Artificial Intelligence, Designer Genomes, Exploration of Ethical and Unethical Research Methods, Medical Practice and Scientific Experimentation

---

How Experimental Objectivity is not as certain as we think

Clare Lee and Miguel Barranco survey forms of experimenter bias and how it manifests to produce a noticeable difference in fields traditionally known for emotional neutrality.

Examination of the definition of Scientists

Michelle Mendez examines the current definition of a "scientist" and argues the term may be applied more broadly to include an actively learning public.

Featured in chapter four!

Take Action

Carmen Conroy challenges preconceived notions concerning lab dynamics and the relationship between mentor and mentee. She provides clear insight into the elimination of bias in workplace and

Social Skills and Science Careers

Julie Mendoza analyzes the context for scientific and academic achievement and questions the merit of an education centered aggressively on empirical

Advances in uncharted territory require that considerations and regulations remain relevant. Oversight can reinforce the line between what is safe and ethical before these advances are made. It is always possible for the application of future technologies to pose a risk. However, like most heavily-weighted decisions, this potential issue requires feedback from multiple perspectives. This may result in slippery slopes with outcomes that could not have been predicted and questions that are sometimes left open-ended. Living in the modern world of evolving cultures with adaptive natures deciding what is appropriate can be stifling. Reaching a consensus could mean abandoning unborn ideas or spending taxing amounts of effort for inconclusive results. What should be understood, with no exceptions, is that no degree of impatience should lead to hasty discussions that lack regard to those who may be potentially affected in the present or future. In the following essays it is agreed that appropriate moral protocols should be available and understood by everyone in order to ensure the greatest amount of safety. Several perspectives will be provided on topics varying from designer babies to medical ethics in hopes of facilitating a deeper conversation regarding the controversies in "Ethical Concerns in Science".

It's been proven that conversation is an effective means of implementing impartiality and diversity within any policy. It is also a way of confronting the invisible consequences that result from advancements in science. To avoid common misconceptions all projects presenting research should include, not only the research itself but, its sociological, anthropological, and technological influences.

In "Bias in Science" we discuss the biases held by and against individuals in laboratories and the public. These biases have the ability to change the way people engage with science in terms of interest and even accessibility. Lack of representation in this field creates assumed exclusivity and brings into question whether institutions are in need of diversity. Any of these potential problems can impact the progress of research that rely on the trust of the people. Dr. Charis Thompson, Associate Director, and the founding Director, of the Science, Technology, and Society Center at UC Berkeley, mentions that the "scientific pipeline" expects individuals to be proficient in science early on in life. However most members of the public lack or are unaware of the resources available to pursue this understanding. The truth is that although the tools to develop scientific interests exist, their availability often remains out of the peripherals of the average person. Instead the aspects that are advertised tend to perpetuate stereotypes and cause bias by revealing only a one dimensional perspective of those who work in laboratories. This can be recognized in films that consistently show only rigid characters and dangerous experiments that are associated with science. The narrowness of a film lens has great influence in the way a person perceives real institutions and their research. Stretching this lens through interdisciplinary relationships in communities would encourage an impartiality for others to consider.

Another way to successfully share the possibilities in research and convey ideas to a broader audience is by choosing the appropriate language in which to communicate. The eyes and ears of the public are valuable assets to problem solving and the growth of new ideas. Connections can be found by considering the integration of culturally competent language in published research or perhaps allowing multiple versions that cater to specific readers without altering any statistical values.

A discussion between students regarding ways to decrease the barriers and stigmas that separate the imaginative contributors we call scientists and the various people who digest their data lead to these proposed adjustments. As individuals, we are all a part of the discussion. The background of anyone in a particular field effects the work that they do. Hence people are more capable of growing and evolving in the direction of positive change than cogs in a machine.

A lab coat implies an expectation in society. Those who wear lab coats also wear the burden of specialized knowledge. However the decisions and execution of this knowledge remain human. It is a collaborative effort in search of an ultimate understanding. Contrary to popular conceptions, the truth is not finite and has underwent change time and again.

To further explore this topic this chapter provides personal narratives from scientists describing the challenging dynamic between a private and public understanding of science in "Taking off the Lab Coat."

Part 1: Ethical Concerns in Science

An exploration the two main issues that come to mind when dealing with scientific experiments: safety and morality.

Diego DuBon: Ethical Concerns in Science: Medical Ethics

Our group's overall topic is "Ethical Concerns in Science" and within this topic we will be focusing on safety and morality within the biological sciences. My own personal focus within this topic will be on Medical Ethics, mainly the ethical considerations taken by doctors on a day to day basis. Overall, what I will be trying to do here is not to determine what are the most moral or right actions to take in these situations, but rather I want to show the thought processes of doctors when they are put into these situations and if they differ at all from how non-healthcare professionals make ethical decisions. In order to do this, I will be considering four common issues within medical ethics within hypothetical cases: euthanasia, organ donation/transplantation, patient consent and confidentiality.

Euthanasia, which translated from its Greek root meaning means "true death", is a term used to describe the act of a doctor willingly killing a patient, with the patient's consent of course. A more formal definition gives us "the painless killing of a patient suffering from an incurable and painful disease or in an irreversible coma" plus "The practice is illegal in most countries", but "As of 2015, euthanasia is legal only in the Netherlands, Belgium, Colombia and Luxembourg" (Google; Legality of Euthanasia). To make things easier for us, we will imagine that we are in a place where euthanasia is legal and we will not be taking into account assisted suicide as it falls into a different category. Obviously, we are still taking into account the Hippocratic Oath, or at least a version of it, since all doctors essentially follow it to some degree. The Hippocratic Oath is an oath that all doctors swear by that essentially states that they will honor a patient's confidentiality, do no harm, and they will not perform any treatments for criminal purposes. Now let us consider two cases, one where a patient is suffering from an incurable and painful disease (let's say an incurable cancer), and one where a patient is in a coma and they will most likely (let's say 99.9%) not wake up. So let us think about how we might justify euthanasia for both cases starting with the former. Now before we can think of even recommending euthanasia, the doctor has to let the patient know what will happen to them that their cancer is incurable and as it progresses, it will be increasingly painful and the doctor will also let the patient know about their options including euthanasia. Even before recommending euthanasia, as it is a last resort, the doctor has to take into account things like the patient's state of mind (they might be suicidal, or overly emotional as they found out they have incurable cancer, etc.) and if there are other options like experimental treatments, alternative treatments, or even just the prospect that the cancer might go away. Although in the second case, it would not be patient's consent and feelings we must worry about, it is the next of kin or whoever has power of attorney since the patient is not conscious. Now one thing the patient and the doctor must take into account is the slim chance (in this case 0.1%) that the patient might wake up, as there is no treatment for a coma. In terms of state of mind, the doctor has to consider not just the emotions of the next of kin, but there motivations as well, since the next of kin or next person with power of attorney might have other motivations for wanting euthanasia (inheritance, money, etc.).

Organ donation/transplantation together essentially are the taking of one person's organs (usually a dead person) and giving them to another person because their organs are failing, damaged, or just generally in poor health and beyond treatment. As an ethical issue organ donation and transplantation have separate but related considerations. One of the easiest ways to acquire organs from someone is that selected to be an organ donor on their driver's license when they were alive and when they died a doctor/hospital has the right to take the person's organs and give them to someone on the transplant list. Where the ethical issue in donation can be found is when the patient has not already selected to be an organ donor but when they die they have healthy organs which could be used for transplantation. The considerations a doctor must consider here are who has the right to "give away" the patient's organs, mainly next of kin or whoever holds power of attorney once the person has died, not to mention that there is also a time constraint if someone needs the organ immediately or there is a "time of expiration" for the organ. From here a doctor must consider the emotions and state of mind of these people before trying to acquire the organs. For the transplant patient the doctor must consider things like, has the patient followed the correct protocol for receiving the organ and do they actually need the organ or is there another way to save them? In this case there is a system set up, however flawed, for a person to receive organs from another. It seems most of the considerations are from taking organs from people.

Patient consent is something I've touched on with the previous topics and is an ethical issue that comes up a lot in medical ethics. This is because when dealing with people's lives, doctors cannot simply make their own decisions in terms of what treatments a patient should receive, mostly because it is someone else's life they are dealing with not their own. One rule of medical and other biology-related ethics is to respect a person's autonomy and one must be very careful when dealing with people who do not have as much autonomy as others like, people in a coma, prisoners, soldiers, etc. Of course this topic becomes tricky when it involves health issues that not only benefit one person, but many people. For example, we have the case of vaccinations; many people opt out of receiving vaccinations or refuse to vaccinate their children because they think it might harm their child and even though doctors know that science has proven otherwise, they cannot simply force people to get vaccinated or to vaccinate their children.

Confidentiality is an issue people don't really think about too much because they think it is an issue that has been resolved. While under normal circumstances, a doctor is supposed to and will honor doctor-patient confidentiality; there are cases when it is violated. For example, if a patient comes in and is dying, the doctor must immediately take into account any precautions to prevent the patient's death like if they are allergic to any medications and when the doctor finds this out they have to let all the medical staff know this so someone doesn't accidently give the patient that medication and since the patient is dying, the doctor does not have the time to ask the patient for consent. Another case is if a doctor is talking about a patient with another doctor when the patient is not because they need a second opinion. Is it okay to give the patient's information then, especially if the other doctor has treated the patient before?

These are only a few considerations for four issues within Medical Ethics. There are many more opinions, considerations and interpretations that I have not mentioned here. But now it possible to see some of the things doctors have to think about every day.

Works Cited:

"Google Definition of Euthanasia." Euthanasia definition - Google Search. N.P., n.d. Web. 07 May 2015. <https://www.google.com/search?q=euthanasia%2Bdefinition&ie=utf 8&oe=utf-8>.

"Legality of Euthanasia." Wikipedia. Wikimedia Foundation, n.d. Web. 07 May 2015. <http://en.wikipedia.org/wiki/Legality_of_euthanasia>.

Joseline Padilla Alvarez: Ethical Aspects of Designer Babies

The technological advances that exist today have allowed us to accomplish things that we'd only though we would see in movies. However, along with these advances and accomplishments, comes a lot of controversy and debates that question how much power do scientist really have? It is almost impossible not to have these debates because everyone has a different opinion and different perspective based on their personal experiences. This can be in the field of genetics where it is now possible to screen fetuses in order to figure out if they are carriers of different diseases, however science has taken this a step further and now offers the possibility to genetically modify fetuses in order to weed out any diseases that can affect the child as it develops. Although having the option to have a "designer baby "that is disease-free sounds great to some, to others prenatal engineering sounds like the modern version of eugenics.

People who argue that genetically engineering babies in order to rid them of diseases is a good idea, believe that this should be mandatory because it would bring great benefits to our society. Some of the benefits that people claim prenatal engineering has are not having to deal with expensive health care later in the future, better genepool, and that parents won't have to put their child through misery.

However, prenatal engineering is now being proposed to be used for so much more than disease prevention, scientist and doctors want to use this method to change the personality of babies in order to have "less harm" in our society. Another idea that has been proposed, is the idea to be able to change a baby's physical features in order to have "better genes". The idea of designing babies, leads many to question if this could be the modern form of eugenics, a way to biologically "purify the human race" and avoid "unwanted people". This can lead to very controversial ethical debates that bring in religion and morals.

People who argue against prenatal engineering, claim that it is unethical because it suggests that the people who are alive now should not be alive and that scientists are now playing the role of God. It seems that science and religion will always have opposing ideas and neither should expect the other to follow each other's practices. Thus why we must continue to respect freedom of religion and not make anything mandatory because this would cause even more serious debates between the two. Heavily debated aspects should be allow people to make a personal choice whether or not they want to become a part of something like designing a baby. If everyone would learn to respect each other's personal choices then a lot of the ethical concerns that exist throughout science could be avoided.

Audio:

http://mcbfifteen.tumblr.com/post/118593000345/ethical-concerns-in-science-audio-by-albert/embed

Infographic:

http://mcbfifteen.tumblr.com/post/118593129830/ethical-concerns-in-science-infographic-by-albert

Joseph Jweinat: Is it Safe and Moral to Edit the Human Genome?

One of the biggest topics in both popular science and bioethics today concerns safety and morality issues when editing the human genome. In an article titled "Scientists Seek Ban on Method of Editing the Human Genome" by Nicholas Wade, a myriad of opinions by different scientists are given concerning these two central issues. Interestingly, while some scientists support the method of editing the human genome and some are against it, they all do agree that it is a dangerous process. Wade perfectly explains why this method is dangerous when he writes, "The chief problem has always been one of accuracy, of editing the DNA at precisely the intended site, since any off-target change could be lethal. Though highly efficient, gene-editing techniques occasionally cut the genome at unintended sites." With safety issues being covered, morality issues can now be studied.

When editing the human genome, scientists have differing opinions when it comes to whether the method is moral or not. On one hand, according to the article, it is moral in the sense that it can help cure genetic diseases. On the other hand, there are two immoral concerns. The first concern is one that has to do with religion. By editing the human genome, many people (even scientists) believe that we are playing the role of God, which is against many religious beliefs. The second concern deals with humanity. If people wish to improve strength, for example, they would edit their genes for more muscles. This would lead future generations to look less human and more like monsters. Such immoral concerns like the ones above should be taken into deep consideration.

Personally, I believe that editing the human genome is both dangerous and immoral. When it comes to editing the human genome, any minor mistake could result in a major disaster. If the genome is cut at unintended sites, the effects could possibly be lethal. Instead of taking the risk of such lethal effects, we can simply avoid them by disregarding the process overall. Although editing the genome can cure genetic diseases, I still believe it to be a highly immoral process for the same reasons claimed by the people within the article. In all honesty, I am a religious person. When it comes to something that is similar to playing the role of God, I am completely against it. God made each human being in His own image. If we cannot accept the beauty that lies within each of us, then what do we see ourselves to truly be? People should be happy with what they are given, since we are all unique in our own way. By editing our genes, we are doing nothing but denying the beauty within ourselves. We are basically claiming that we are not happy with ourselves and with what God gave us. In my view, this is highly immoral and unacceptable. Transitioning from a religious perspective to a social one, I view gene-editing techniques to impact humanity in a negative way. The reasoning for my perspective is exactly the same as the example above that deals with strength. Now that I have covered my own personal beliefs about this process, let us focus on this issue in a more general perspective.

If the scientists involved in editing the human genome would consider my perspective above, then a numerous amount of problems could be avoided. As mentioned before, lethal effects could be avoided if the process is disregarded overall. To add on to this perspective, there surely must be another (undiscovered) way to cure genetic diseases. With technology expanding and improving at such a quick rate within today's time, a new, safe way to cure genetic diseases will soon be discovered that will avoid any lethal effects. Moving on, no religious beliefs will be contested against if the process is disregarded. Finally, people can still be considered as humane rather than monstrous if we avoid such things as editing human strength.

Overall, there does not appear to be any good reason besides curing diseases that show why editing the human genome is a marvelous process. Rather, a myriad of thoughtful points concerning safety, religion, and humanity have been brought up when dealing with this issue. In my opinion, since there are more negative aspects than positive ones, this process should be disregarded overall. By considering my perspective, both humanity and society can continue to flourish as wholes. In addition, disregarding this process will help bridge the gap between science and the public. Everyone is mixed in their opinions when it comes to editing the human genome. If we just disregard this process overall, there will be no mixed or conflicting opinions. Thus, we can move one step closer towards bridging the gap.

Group Project URLs:

http://mcbfifteen.tumblr.com/image/118593129830

http://mcbfifteen.tumblr.com/post/118593129830/ethical-concerns-in-science-infographic-by-albert

http://mcbfifteen.tumblr.com/post/118593000345/ethical-concerns-in-science-audio-by-albert/embed

Works Cited:

Wade, Nicholas. "Scientists Seek Ban on Method of Editing the Human Genome." *New York Times* 19 March 2015: Print.

Jazmine Carvajal: Artificial Insemination: Is it Acceptable?

Imagine a situation in which you and your partner had always desired to have a child but were unable to do so naturally. How would you go about this situation? Would you consider alternative methods? Infertility could be a difficult situation to cope with, especially if a couple's desire is to form a family of their own. Artificial insemination could be an option. This alternative type of fertilization is not only used as a resort when someone is infertile; it is also used when a woman would like to get pregnant without having a male partner or when a sperm donor is required. Because of this, many individuals resort to artificial insemination (AI) as a way to fulfill their dream of forming a family. However, some individuals might be against AI due to their religious views or beliefs.

Ethical and moral concerns regarding artificial insemination are put to question when a sperm donor is needed to go through the process. Many people, who deeply follow their religion, argue that having another person's child out of their marriage or relationship is not acceptable. Morality issues arise because many religions promote conception between two individuals, not three. Concerns regarding the child's parents are also brought up when considering who provided their genetic material to conceive the child and what "rights" they have over the child. If a donor would be involved in the process, a question regarding their role in the child's life is put into play. Would they be allowed to meet the baby or have a special role in their life, would it be okay to take the donor's right as a parent from the child? The answers to these types of questions vary from state to state, depending on the laws that are enforced within different regions.

There are also safety concerns regarding the process of AI. Even though, AI is meant to help a woman give birth to a child, there is a possibility of the transmittance of unforeseen genetic diseases. According to Ashok Agarwal and Shyam Allamaneni, artificial insemination could also harm the woman carrying the child. The mother could suffer pelvic infections, allergic reactions, or have pregnancy complications, such as multiple pregnancies at once or have a "spontaneous abortion". Therefore, one can see how AI can be dangerous but at the same time may be a positive thing. Dr.

Gerald Perkoff published a research article of a lesbian couple that was not successful in conceiving a child through the process of artificial insemination. The couple made special requests when deciding whom to choose as their sperm donor because they wanted their child to resemble similar racial characteristics. Having gone through many cycles of insemination, one of the partners experienced difficulties bearing the child, which led her to have an exploratory laparotomy and later on an abortion. After others learned about this case, articles that questioned AI in lesbians were written for newspapers. After receiving harsh judgment due to the morality of the issue, the couple no longer decided to continue with the process. The couple's physician also experienced difficulties and harsh criticism for trying to help the couple give birth to a child. This example exemplifies the moral, ethical, and safety issues that comprise artificial insemination as a whole.

Overall, artificial insemination is a topic that should not be considered lightly. Every aspect should be researched before making a final decision whether to follow through the process or not. Couples who consider going through this process should investigate the complications that they might encounter and any family history of genetic diseases. Artificial insemination can either be the best thing in a couple's life that can bring them happiness or the most unfortunate. Although it may raise and question morality and safety, one must become educated before making opinions on the subject.

Group's Media

Info graphic: http://mcbfifteen.tumblr.com/image/118593129830

http://mcbfifteen.tumblr.com/post/118593129830/ethical-concerns-in-science-infographic-by-albert

Audio: http://mcbfifteen.tumblr.com/post/118593000345/ethical-concerns-in-science-audio-by-albert/embed

Anonymous: Prison Experimentation

Throughout the 20th century, there were huge uproars among scientists and the mode of which they had carried out their studies. From the horrendous war crimes by the Nazi's medical researchers, to the awful injection of live cancer cells in a hospital at New York, prison experimentation has been a threat to both the morality and humanity of the prisoners and our basic human rights. One notable case, called the Stateville Penitentiary Malaria Study, had been testing the effects of malaria on many prisoners in the Stateville Penitentiary near Joliet, Illinois.

Heading the research, the medical department of the University of Chicago and the United States Army had injected many mosquitoes with a strain of malaria and over hundreds of prisoners were injected with malaria and were studied of symptoms and possible cures. Oddly enough, after hearing about this process, a large proportion of the people had been rather indifferent to this type of testing because of the incredible benefits that this study had brought about, including a better knowledge of malaria and potential cures toward this disease. However, after several lawsuits were filed against these experiments, the case went to the Nuremberg Medical Trials, also known as Doctors' Trial, which was a result of the inhumane experiments on people by doctors of the German Nazi regime. The defense attorney had argued that there were no difference between experimentation done in Nazi concentration camps and American prison grounds and that it was ethically wrong to perform such experimentations which pose a threatening risk to its subject. Ultimately, the experimentation was justified by the ideal conditions and ethical standards that were upheld in the experiments. With this study, one of the biggest changes to human experiments was the development of the Nuremberg Code which holds the principles and ethical standards of which human experimentation may take place and under the Nuremberg Code, the first protection to the ethics and safety of experimentation was created in order to help protect researchers and subjects alike from the inhumane potential of medical experimentation.

In order to provide autonomy to the prisoners which are subjected to these experiments, it is highly important that we take measured steps to help educate these prisoners of the risk and make sure that there are no threats, incentives, or other means of coercion placed upon these subjects in order to provide them with the best possible sample of their situation and what is to come. These laws and codes established all work to help fulfill these tasks in order to treat all human life equally and fairly, and to help create a healthy, moral, and safe experiment.

Taking an overarching look at prison experimentation and the safety and morality of humanity, there is an extremely thin line between what is considered appropriate and what is considered unethical. However, through strict regulations by scientists and the government, there have been huge strides taken in order to help improve the quality, success, and efficiency of both human rights campaigns and advances in the fields of medicine and biological studies. As long as we continue to advance our knowledge in current controversial topics in research, such as abortion, human experimentation, vaccination, etc., we must keep in mind the importance in establishing a healthy and safe environment for all participants in these studies and maintain a morally upright and justified mindset throughout the process. Prison experiment heavily relies on the consent of the individuals, and thus a safety net for these individuals must be established in order for there to be a successful and ethically just experiment.

Terry Kyubin Kyung: Moral Dilemma around Embryonic Stem Cell Research

My first exposure to stem cell research was when professor Hwang Woo-suk, a publicly renowned South Korean veterinarian and researcher, was accused of fabricating his experiments in stem cell research. With his scandal the controversy of morality of stem cell research surfaced in Korea. Coming from a Christian background, this issue was not a dilemma for me. Because of my faith, I believed that life started as soon and the egg is fertilized and I could not understand why people were arguing about when the life actually starts. Later when I entered college, taking science classes and finding more about stem cells, I understood why the issue of stem cell research was controversial and that it was not an easy question to answer. I also realized how it was so easy to take a side and argue blindly without necessary information to consider.

When studying stem cells, there are two kinds of stem cells. One of the two types is adult or "multipotent" stem cell. This kind of stem cells can regenerate only one type of cells such as blood cells regenerating at bone marrow and skin cells regenerating after sunburn. Research in this kind of stem cells does not raise ethical issues because the adult stem cells is a part of one's body and not a person. On the other hand, embryonic stem cells, also known as "pluripotent" stem cells, can generate into any part of the body, which means that these pluripotent cells can grow into a full human being. The human embryonic stem cell research (HESC) generates moral dilemma because people think differently on at which stage is the fertilized egg considered a live human being.

The opponents of embryonic stem cell research state that development from a fertilized egg into to a baby is a "continuous process and any attempt to pinpoint when personhood begins is arbitrary," (EuroStemCell). They also state that if there is uncertainty whether a fertilized egg should be considered a human being, then should not destroy it just as "a hunter does not shoot if he is not sure whether his target is a deer or a man" (EuroStemCell). On the other hand, the supporters of embryonic stem cell research state that fertilized eggs younger than 14 days cannot be given personhood. About a third of fertilized egg die naturally from failing to implant on uterus wall. "If the natural process involves such loss, then using some embryos in stem cell research should not worry us either," (EuroStemCell).

This moral dilemma has divided the international community and even among different religions. In the European Union, human embryonic stem cell research is permitted in Sweden, Finland, Belgium, Greece, Britain, Denmark and the Netherlands; however it is illegal in Germany, Austria, Ireland, Italy, and Portugal. In the United States, several states enforce a complete ban and others give financial support. Elsewhere, Japan, India, Iran, Israel, South Korea, and China are supportive of HESC research. Australia is partially supportive; however New Zealand, most of Africa (except South Africa) and most of South America (except Brazil) are restrictive (Wikipedia). Roman Catholic, Orthodox and conservative Protestant Churches believe the embryo has the status of a human from conception and no embryo research should be permitted. In contrast, Judaism and Islam emphasize the importance of helping others and argue that the embryo does not have full human status before 40 days, permitting some research on embryos.

The question of when the fertilized egg is considered a human being is important because the process of extracting desired embryonic stem cells kills the embryo. If the embryo is considered a human being at the time of extraction, then the process becomes an act of killing human life.

Despite the moral dilemma it faces, the embryonic stem cell research is still regarded as an important study for discovery of new medical treatments for diseases and disabilities. In an attempt to evade the moral dilemma altogether, the researchers developed an alternative, called induced pluripotent stem cells (iPSCs), which is a genetically "reprogrammed" adult stem cell to function like an embryonic stem cell. Again, adult stem cells are part of a body and do not cause the moral dilemma of destroying life. The iPSC could completely replace human embryonic stem cells and eliminate the controversy around embryonic stem cell research. However, the study on iPSC is incomplete and we do not yet know whether iPSC could completely replace embryonic stem cells still and whether we can safely transplant them into humans (SEP).

Depending on their culture, religion, and interpretation of scientific facts, the people react differently to the controversy, differing in opinions about when the life starts and whether human embryonic stem cell research is ethically acceptable. Educating the non-scientist people on the science and mechanism of embryonic stem cell may not necessarily subdue the controversy, but it may motivate them to ponder about the ways to resolve the controversy. Also, through the development of iPSC, the scientist have demonstrated that finding a new approach that eliminates the very question that causes the controversy, rather than trying to find a middle ground between two opposing sides, can be the best solution to a controversial issue.

Works Cited:

1. "Embryonic Stem Cell Research: An Ethical Dilemma." EuroStemCell. N.p., n.d. Web. 11 May 2015.

2. "Stem Cell Research Policy." Wikipedia. Wikimedia Foundation, n.d. Web. 11 May 2015.

3. Siegel, Andrew. "Ethics of Stem Cell Research." Stanford University. Stanford University, 25 Apr. 2008. Web. 11 May 2015.

Denisse Velazquez: Harm Reduction: A Public Health Strategy

Harm reduction acknowledges the inevitable presence of drug use and does not attempt to end it, but rather reduce its harmful effects on users and on their relationships with others, or with their environment. Does this actually work? By providing supervised injection facilities, it diminishes the probabilities of overdosing and transmission of diseases such as HIV and viral hepatitis. There is a huge controversy on how harm reduction affects health, social and economic aspects. However, when focusing on health outcomes in a population where abstaining from drug usage is not realistic, there is a clear advantage.

For a long time drug abuse has caused many deaths due to overdose and disease transmission. Supervised injection facilities provide a space where safety issues are targeted by providing sterile syringes, needles, cotton, and strings, among other supplies as a strategy for reducing risky health behaviors among users. Risky health behaviors are an issue because HIV and viral hepatitis are an immense health concern. Harm reduction programs have proven to be effective through programs such as the Portugal model where they have the lowest rate of HIV in Europe. According to the article "Treating, not punishing," Portugal decriminalized the personal use and procession of drugs and this this led to Portugal's HIV cases relating to drug use to decrease, "Drug addicts now account for only 20% of Portugal's HIV cases, down from 56% before." The Portugal model is a great example of the effectiveness that harm reduction programs has in the health aspect and proves how the reproduction of these programs can have a positive effect in a society.

The bay area is fortunate to have people who saw the harm that unsafe usage of syringes and needles amongst drug users made in communities where drug abuse was prevalent. When referring to the San Francisco Aids Foundation website, the Prevention Point program began needle exchange in 1988 in San Francisco, providing clean syringes for the exchange of dirty ones. This program began with a group of volunteers that understood that providing a safe way to users to inject themselves, also served as a way to educate the users on how to prevent disease transmission and overdose. This program has become very successful and improved the health of an entire community of drug users. Berkeley has also replicated these strategies through the Berkeley Needle Exchange Emergency Distribution where overdose prevention is targeted through the distribution of naloxone, which is reverses opiate overdoses.

Ultimately, even though lives are saved through disease prevention and overdose prevention, controversial questions still remain concerning issues with harm reduction such as: are these facilities promoting drug use and thus sending the wrong message or is it unethical to provide no assistance for drug users which leads to disease transmission and high mortality rates? How do we draw the line between what is ethically correct and safe? What role should the medical community have in these injection facilities? Will effects become negative in the long run? In the end, harm reduction programs practice safe ways in which people take their drugs, and at the same time I believe that it is a way to educate drug abusers on how to reduce their health risks, creating a healthier society.

Works Cited:

Greenwald, Glenn. (2009, August 27) *Portugal's drug policy: treating, not punishing.* Retrieved from http://www.economist.com/node/14309861

"History of Health: Needle Exchange in San Francisco." *San Francisco Aids Foundation.* Web. 01 May. 2015. http://sfaf.org/client-services/syringe-access/history-of-needle-exchange.html

Part 2: Bias within Science

A wide ranging representation of consequences, forms, and manifestations of bias within the scientific community. This includes, but is not limited to, scientists who are and have been discriminated against, bias within the process of conducting research, and the overall result this bias has had on the public and the scientific community, and the actions one may personally take to discourage this bias from continuing.

Sabrina Rentschler: Unsung Heroes in Science

Throughout history many great scientists overcame societal prejudices to provide mankind with the most important and pervasive technological advancements known to man.

African American scientist, Percy Julian, was born in deeply segregated Montgomery, Alabama; although this prevented Julian from attending public school past eighth grade, he persevered to pursue Chemistry at De Pauw University. The university's racist policies did not allow him to live on campus and dining halls refused to serve him meals. Eventually, a fraternity allowed him to stay in their attic in exchange for housework. He would later manipulate sterol taken from soy beans to chemically create sex hormones testosterone and progesterol. This breakthrough greatly improved pre-natal care and made chemical birth control possible. He also manipulated soy beans to improve the techniques to make cortisol, causing it to be produced at a fraction of its previous cost and effort, greatly reducing the expense of organ transplantation and the treatment of autoimmune diseases.

Alan Turing made the single most significant contribution to the allied victory in World War II when he created a functional model for the general purpose computer to break Nazi codes. His breakthrough allowed allied forces to decode Nazi messages and shorten the war by several years. However, after founding principles central to artificial intelligence and modern computing, Turing was outed as a homosexual and forced to endure chemical castration ordered by the British government in order to avoid prison, prompting his suicide.

Hedy Lamarr escaped Nazi controlled territory and an abusive marriage to a fascist weapons manufacturer by disguising herself as a maid; she escaped to Paris, and later America. She later became a famous actress and was known as "the world's most beautiful woman". Lamarr successfully created and patented a new technology she called "frequency hopping", meant to protect American torpedoes in World War II. Her technology was far ahead of its time, but was never used during the war; she was instead encouraged to use her sex symbol status to sell war bonds. Her ideas were ignored nearly twenty years until used by the American military in the Cuban

Missile Crisis, and more recently as the heart of wireless technologies we use today, namely cell phones and blue tooth technology.

After graduating college, Henrietta Swan Leavitt studied stellar spectra through photographic plates in an all-female research team at Harvard College Observatory. As a research assistant, or "computer" as they were called, she was not allowed to use a telescope; women were forbidden from using the world's largest telescopes until the 1960's [1] In 1912, although she had gone deaf and had been paid poorly or not at all, she discovered the Period-Luminosity relationship, becoming the first to create a substantial method of measuring stars' distances from the earth. At this time scientists could only measure 100 light years into space, whereas her discovery permitted a breadth of ten million light years into space[2]. She was later forced to change her research focus by her supervisor and was not able to continue this research; however this law became key in measuring distances in space; it allowed for the Hubble telescope and the discovery our expanding universe; it allowed for many of the aerospace achievements of the 20th Century.

African American physician, Dr. Charles Drew, revolutionized medicine by separating plasma from whole blood to keep blood viable for a longer period of time, in doing so, he greatly lessened the blood's likelihood of contamination and innovated the world's first blood bank. By creating the first bloodmobiles he encouraged blood donation as part of the World War II effort, helping build blood bank foundations as part of the American Red Cross. Dr. Drew spoke avidly on discrimination and would eventually quit his job with the Red Cross in protest of racial segregation of blood used in transfusions; however he continued to make progress in transfusion methods, surgery, and education. His methods have saved countless millions of lives.

Vera Rubin graduated as the only astronomy major at Vassar, a women's college. She attempted to join Princeton's graduate Astronomy program, however this program did not accept women until 1975[3]. She went on to study at Cornell as a student of Richard Fenyman. Vera Rubin was the first to describe the "galaxy rotation problem", or the discrepancies within the predicted movements of galactic curves. Her data became the first to legitimize the existence of dark matter. Although she is still living she has yet to win the Nobel Prize.

Ada Lovelace was the first to create what she called "the abstract science of operations", now commonly known as computer programming. In 1840, she was the first to write an algorithm for Charles Babbage's Analytical Machine, the first theoretical calculator. She was also the first to theorize the programmable modern computer by insisting that the machine could be more than just a calculator. Although Lovelace's understandings surpassed Babbage's own, he thought of her as an enthusiastic beginner and she was never allowed to work for him formally. When providing her contributions in the form of notes, or submitting papers for publication, she used her initials instead of her name to hide her gender. Her work was ignored until it was rediscovered by Alan Turing. The suppression of her thoughts and ideas is believed to have delayed the computer age by a hundred years.

In light of these achievements and others, it is evident what great respect is owed to these individuals for their sacrifice, brilliance and patience, and the imperativeness of avoiding limitation of potential innovators in the future; a great scientist can come from anywhere to achieve anything.

If you are interested in more topic concerning bias in science, please access the infographic at http://mcbfifteen.tumblr.com/post/118649952600/infographic-for-chapter-4-scientists-are-people

If you are interested in the infographic corresponding to this presentation it can be found at https://magic.piktochart.com/output/6040435-untitled-presentation

Works Cited:

[1] http://w.astro.berkeley.edu/~gmarcy/women/history.html

[2] http://physics.weber.edu/carroll/fivewomen/5women.pdf

[3] http://physics.weber.edu/carroll/fivewomen/5women.pdf

Catherine Link: Underrepresented Minorities in STEM

In more recent years, there has been a slight increase in the proportion of historically underrepresented minority groups in science and engineering majors. According to the article "Minorities, Women Still Underrepresented in STEM Fields,

Study Finds" written by Alan Neuhauser from *U.S. News*, the percentage of racial minorities in the STEM field has shifted from 7 percent in 1993 to 10 percent in 2010 and the percentage of women, being another underrepresented minority, has shifted from 21 percent of STEM majors to 28 percent in 2010. This is evidence that the scientific field is diversifying, but it also demonstrates that STEM is not keeping up with the diversification of the general population. Not only have the numbers of ethnic and female STEM majors decreased, but so has the percentage of those who continue in those fields of study, with only 31 percent of minorities graduating with STEM degrees continuing in their intended fields.

A majority of the world's population belongs to a historically underrepresented ethnic group, yet the bulk of research conducted to better the world is being done by the minority: white males. However, in science, the white male population is considered the majority and the outliers make up the minority. This lack of diversity impacts the effectiveness of scientific research because those conducting said research cannot be completely cognizant of the problems the people their research is directed at face. Charles Lu, director of academic advancement and innovation at the University of Texas at Austin, was quoted by Andrew Jones in *The Brown Daily Journal* saying, "You have a big change in demographics, yet our people of color are not getting into STEM fields, or they're getting pushed out. And that poses a big problem, not only for the education system but also down the road ... Science relies on individuals with diverse perspectives working together" and this requires a certain empathy with their subjects and research to make the scientist more responsive to those demographics. Making the science community more exclusive, makes their research that much less affective, which can have negative consequences in the future.

This lack of representation, not only affects the populations being served, but it also deprives students and professionals of sharing their talents and creating growth in their respective fields. Natalia Chabebe, a Teach for America STEM teacher, shared her experience of being a Latina in the engineering field with *The Huffington Post* and wrote "lack of diversity in STEM is dangerous. It's what makes teachers and students doubt what is possible ... [my] experiences allowed me to follow my passion, but they also drove home how rare it was to be a young Latina in engineering, and how necessary it was for there to be more diversity in the STEM fields." Chabebe was raised in Ecuador and throughout her life, both as a student and as an engineer, people have always dissuaded her from continuing in STEM because they didn't think she could make it because she didn't meet the standard. She now advocates for even distribution of resources to all demographics so that, as she put it, "no one will ever question their abilities because of how they look and a lack of precedent." Because these groups have been discounted throughout the years, their expertise has been undermined and this has further dissuaded these groups from the STEM community.

Chabebe, Natalia. "Lack of Diversity in STEM is Dangerous for our Students." <www.huffingtonpost.com> April 26, 2015.

Jones, Andrew. "Minority groups underrepresented in STEM Fields." <www.browndailyherald.com> April 26, 2015.

Neuhauser, Alan. "Minorities, Women Still Underrepresented in STEM Fields, Study Finds." <www.usnews.com> April 26, 2015.

Infographic: http://mcbfifteen.tumblr.com/post/118649952600/infographic-for-chapter-4-scientists-are-people

Audio File: http://mcbfifteen.tumblr.com/post/118666764285/this-audio-file-delves-into-the-sources-of-biases

Or https://soundcloud.com/adm-buttercrust/mcb-biases-within-science

Music in Audio file is "In Albany New York" by the 126ers from the YouTube Audio Library

Claire Lee: Bias in the Process of Research

Many people may expect science to be a very objective field of study, due to the meticulous steps that must be taken to gather and analyze data. However, experimental error or failure to consider all possible variables may introduce bias in research and science. Like in many other disciplines, bias is unavoidable. Thus, it is important to recognize the potential biases that may skew the results, and take extra care to avoid such biases as much as possible.

The most common form of bias is the Experimenter bias, which is a type of bias in which researchers may influence the outcome of the data to get the results that they want. This type of bias is very difficult to prevent, because it stems from the inability for humans to be totally objective about a subject. This is especially true if the experimenter has a hypothesis that he or she is testing, and may subconsciously be looking for evidence or data to support that hypothesis.

A 1962 twin study of intelligence conducted by James Sheilds illustrates how experimenter bias can influence the data collected. Sheilds tested the difference in IQ scores of 44 identical twins that had been had been raised apart. When Sheilds analyzed the scores himself, he found the average difference to be 8.5 points. However, when other independent examiners tested the twins, they found the average difference in IQ scores to be 22.4 points – a significantly larger number. It has been suggested that this difference in analysis could be due to the fact that Sheilds was biased, expecting the identical twins to have similar test scores, while the other experimenters were able to more objectively analyze the IQ scores. [1]

One method for combating experimenter bias would be to employ double blind techniques when designing studies to ensure impartiality. If neither the groups being studied nor the researchers know who is placed in the control or experimental group, it would greatly reduce the chance of bias.

Another type of bias is measurement bias. In any type of experiment, human error in measurement is always possible. Though scientists try their best to make exact measurements and calculations to obtain the most accurate data, even little mistakes can be detrimental in data analysis. This can be prevented by using standardized instruments, making sure devices are calibrated, and by being meticulous with taking notes or writing down results.

The best way to perform an objective study would be to consider how biases might occur at which steps of the experiment, prior to execution. Doing so will allow researchers to be more cautious of potential mistakes, and they can take extra care and measures to minimize biases. Although it is impossible to completely eliminate all bias, it is crucial to be aware of how biases may affect outcomes of experiments, and consider how we can reduce bias and study science in a more objective manner.

Audio: http://mcbfifteen.tumblr.com/post/118666764285/this-audio-file-delves-into-the-sources-of-biases

Infographic: http://mcbfifteen.tumblr.com/post/118649952600/infographic-for-chapter-4-scientists-are-people

Miguel Barranco: Stereotypes in the Scientific World

Science is a systematic study of the structure and behavior of the physical and natural world. However, this branch of knowledge is not as impartial as one would think it would be. Interestingly enough, if one looks close enough one could see the multiple biases in the scientific world. My final project is on the stereotypes that are clearly present in the science community and as a result contribute to the partiality in all sciences. Specifically, I would like to focus on the stereotypes of scientists and the stereotypes of the public, on matters of science. In doing so, I hope to understand why the public and scientists have certain beliefs that influence the bias in science.

An example of this could be the stereotypical belief that the public simply cannot understand, and therefore be against publicizing their science. As a result of this, science is seen as a privilege that is only accessible to those who have the knowledge to understand it, scientists. However, the irony in this is that one can never be fully knowledgeable in any specific area of science, there may be people that are better informed, which are scientists but that is only because they have made science bias to them. For Daniel Sarewitz bias is an over-reporting of false positive results, which result from "a powerful cultural belief ... that progress in science means the continual production of positive findings" (Hayton). What this is implying is that scientists have this innate belief that science revolves around them and they have the right to access everything. This is because of the cultural belief that scientist always have the answer, and they can always back it up with data or solid evidence. This in turn branches to a whole other side of bias in science which is confirmation bias.

Confirmation bias occurs when people actively search for and favor information or evidence that confirms their preconceptions or hypotheses while ignoring or slighting adverse evidence. "It is a type of cognitive bias (pattern of deviation in judgment that occurs in particular situations · leading to perceptual distortion, inaccurate judgment, or illogical interpretation) and represents an error of inductive inference toward confirmation of the hypothesis under study." What this tells the general public is scientists have become so focused on obtaining data or evidence that is quantitative to prove their research. As a result many scientists feel the need to manipulate their evidence in order to confirm their beliefs, and this is only done to get in return the necessary funds to carry on their research. However, what this also reveals is that scientist don't necessarily have all the correct quantitative answers, but it is never questioned by anyone in the public because they are "scientists." Nevertheless the outcome of this is that scientists become so obsessed with being the first to obtain the data or make the discovery that they forget science has always been a means to better all of society.

While it may seem that scientists are responsible for a majority of the bias in science, they are not the only one who contribute to the stereotypes, the public does too. The stereotypical scientist is generally described as a white man with lab coat performing research in a lab with chemicals. What is the reason for the public having this stereotypical belief? A large part of the influence is due to the social media but as well as our culture. The way our culture and media has depicted science or scientists has been very tainted because it makes science seem very black and white. When in all actuality it is not, but the public fails to see this because of their preconceived beliefs. As a result it is quite clear that the public is very misinformed, that is not to say that they are ignorant but they simply have not been provided with the appropriate information to understand science. The public only has a small understanding of science or a scientist since their comprehension comes from the social media, and thus explaining why the public's perception of science may be bias. The affects this has on the public is that they form this predetermined impression that they know everything or simply put on this façade that they know science because they overheard it in the news. This results in a lack of appreciation of the actual science is this all due to the fact that stereotypes overpower the actual truth, but it is also because scientists, or anyone from the science community, do not take the time to address the issue.

In the end, this culminates a never ending cycle between the public and the science community because both contribute their fair share of bias in science through their stereotypical views on one another. The solution or solutions for this current issue are not very clear, but the important thing is to recognize that the issue exists and from there small steps can be taken to promote change.

Links:
http://mcbfifteen.tumblr.com/post/118649952600/infographic-for-chapter-4-scientists-are-people https://soundcloud.com/adm-buttercrust/mcb-biases-within-science
http://mcbfifteen.tumblr.com/post/118666764285/this-audio-file-delves-into-the-sources-of-biases

Works Cited:

Hayton, Darin. "Bias Affects All Scientific Research." Darin Hayton (Historian of Science). 14 May 2012. Web. 2 May 2015.

Walker, Michael. "The Deadly Data Science Sin of Confirmation Bias." Data Science Central. 24 Apr. 2014. Web. 1 May 2015. .

Carmen Conroy: Overcoming Barriers in the Scientific Workplace

This section of Chapter 4 will provide readers with a list of concrete resources that they may seek out and employ the use of in order to combat the various barriers, biases, and forms of discrimination that will be characterized by my group members' descriptions. This section will take the form of a narrative which conversationally discusses solutions to problems caused by regularly perpetuated scientific biases, providing anecdotal and institution-based resources to address said issues.

This year, I went to a scientific conference for minority students where I attended a series of workshops designed to address students' concerns about applying for funding and securing research positions, among other topics. In one workshop, we discussed the rotation process that many graduate students undergo to select a thesis lab. We were taught all about positive qualities to look for in a research setting, such as a supportive mentor and PI, respectful peers, and basic courtesy from our collaborators and fellow PhD-ers. As the PowerPoint slides came to a halt and the presenter offered to take questions, a girl raised her hand: "You've talked a lot about what we should look for, like the good things. But how do we know if we're in a bad lab that doesn't fit our needs, or if we have a bad mentor? Like what are things that we should not have to put up with as researchers?"

I immediately raised by hand and started snapping furiously. This is such a critical question to ask, in any workplace. When people fail to ask and consequently answer this question, it usually remains assumed that we—researchers who many times feel subordinate to our PIs and post-docs in the laboratory—should deal quietly with any obstacles thrown our way. However, this is an oversimplified and very dangerous attitude to have. Rather, we should seek to characterize issues presented to us in the laboratory as a result of inherent biases and discriminatory behavior, and then to either extract ourselves from potentially uncomfortable or unnecessarily unsupportive environments, or otherwise choose to combat the situation to the best of our abilities, head on.

I listened carefully to the answer provided to the inquisitive girl's question, cataloging the answers in my head for permanent storage. I will share them with you along with some of my own personal additions. I am sure that there are, unfortunately, many biases that we have to address as a scientific community beyond those which I will present here. However, those I have actively contemplated are as follows:

· Any form of harassment. Many scientists still believe that sexual, racial, religious, ethnic, and personal handicaps are grounds for discrimination and other forms of harassment—crude jokes, unsolicited commentary, and insensitive remarks about your personal ideologies and demographics you ascribe to. You do not need to deal with this.

· Disrespect for your time and needs. While you are a student, your boss or mentor should not expect or ask that you skip class, doctor's appointments, office hours and review sessions, or any other important engagements that you need to and absolutely should attend as planned. If you have a set schedule, stick to it. You are not required to work hours that you do not receive monetary or professional recognition for. You are not obligated to respond to 5 am phone calls requesting additional data. You have a life outside of the laboratory that is altogether yours and needs to be recognized as such.

· Disinterest in your professional and academic advancement. Your mentors should want the best for you. If you come to them asking for a letter of recommendation or express concerns about applying to a conference/lab interfering with academic progress/questions about your future educational endeavors and they are dismissive, they are not fulfilling all of their duties as your mentor. You deserve better.

There are many ways you can address these issues, or, in a worst case scenario, prepare an exit strategy to totally eliminate them. In no particular order, here are some which are readily available specifically to UC Berkeley students, both graduate and undergraduate.

· The Ombuds Office. This is a little known resource which is available at no cost to students, and which provides a totally impartial, confidential, and immediate source of feedback for any issues you experience on campus. The Ombuds Office will provide you with a person on staff who is readily available to listen to the issues you present, and then give you some options and resources that you can harness to address said issue.

o Visit: http://sa.berkeley.edu/ombuds
o Call: 510-642-5754

· The Tang Center. The Tang Center can provide you with group support, individual and group counseling, and general self-help resources that may be of use to you when confronting the damages caused by a biased, unsupportive learning and working environment. Also, it is helpful to note that regardless of insurance plan, any UC

Berkeley student is granted five free independent psychological counseling sessions every academic year.

o Visit: https://uhs.berkeley.edu/students/counseling/cps.shtml

o Call: 510-642-9494

· Direct conversation and open dialogue with the people causing the problem. This can be difficult, but it may be the most obvious first step. Talk to your mentors and other relevant individuals about what you feel is bothering you and any obstacles they have presented you with. Include recommendations for how they could improve their behavior, and keep the conversation more mutual by suggesting ways you could help address the issue as well. Do not apologize for anything that is your fault, and do not pursue further conversation if you continue to be treated in a hostile, demeaning, or condescending fashion.

Colin Wang: The Importance of Universal Vocabulary

Sometimes bias is completely unintentional. In science this often occurs simply because of discrepancies in language. The scientific community has certain definitions associated with specific words, and when the public does not use these same definitions, bias is created unintentionally. This happens with words that are only defined within science as well as words that have different meanings in the public and in the scientific community. Unfortunately, this leads to many misunderstandings between the scientific community and the public. My goal for this project is to help "translate" the language between the two parties, which will hopefully lead to a better understanding of each other.

One of the most commonly used words in science is hypothesis. A hypothesis is key to answering almost any question in science. However if you ask someone from the public what their definition of a hypothesis is, it is likely going to be a different answer than one you would get from a scientist. In fact, even within the academic world, the meaning of hypothesis varies depending on the field of the study. Commonly it is this type of misunderstanding of terminology that leads to communication barriers. This is very evident in the usage of the word theory. I have far too often heard people from the public dismiss theories simply, because they are theories. Many of these people do not realize things that are very fundamental to our understanding of the world such as gravity are what they would consider "just theories" as well. When these communication barriers exist so prominently, it is no surprise that the public and the scientific community don't see eye to eye on many issues.

Other language barriers exists between the scientific community and the general public, when certain scientific don't even exist in the general public. When experiments are being done in labs where everyone is familiar with the lab jargon, this is not as much of an issue; however, when the public is involved, a lot of important information may be left out. The area where this lack of understanding is most effected is in terms of published papers. In the opinion of many scientific journals are the place to get the best most reliable information. Unfortunately, when the public is able to access this information, they might not be able to understand what is going on in research due to the specific scientific language used in the report. If the scientific community really wants to help bridge the gap between them and the public, using language that anyone could understand in lab reports would be a good start. It is true that it will take more effort to translate the specific terminologies; however, it would really help give the public a better understanding of what is happening within science.

It is definitely true that there is a lot that both sides could do to help bridge the gap between the scientific community and the public, however when it comes to language, most of the action should come from the side of the scientific community. Better communication would contribute heavily towards eliminating biases as it would minimize unnecessary misunderstandings. Fixing the language barriers will not solve the entire problem, but it is definitely a significant factor in terms of closing the gap.

Part 3: Taking off the Lab Coat

In addition to exploring the origins of public perceptions regarding scientists, we want to examine scientists as individuals with personal narratives in order to bring insight to how their journey has impacted their relationship with science and what the intentions are for moving from a public to a private understanding.

Julie Mendoza: Prerequisites for Gaining Public Approval

The nature of research is collaborative. It is not the work of a single entity, though science can often appear that way. Individuals collectively execute these studies and make judgments that reflect who they are as people. The men and women in lab coats who are blanketed by the term "researchers" carry histories that influence their work. Their tertiary education is often a precursor for the various fields they find themselves conducting in.

Educational background serves as a way for people to legitimize research. Credentials required to validate the work of a scientist tend to emphasize working within the realm of logic. The framework for true data implies that the individuals experimenting are experts at being first and foremost rational and linear. For example, common requirements for students interested in pursuing a career in science are biology, chemistry, calculus, organic chemistry, and calculus-based physics. Before subjecting themselves to this rigorous academic journey these individuals are not unlike most people in their community. It is their specialized knowledge that begins to differentiate them as scientists. It is also a curriculum that does not necessarily have high value in finding ways to communicate with members of society that lack the same access to specialized knowledge.

To have a positive impact on a community it is fundamentally important to learn how to have constructive dialogue and problem-solve with those who are not necessarily like-minded. Perhaps this intense emphasis on numbers and formula oriented curriculum unintentionally dissolves the will to further develop people skills.

All professions that potentially effect the public require an ability to speak effectively to the many kinds of people. Popular politicians gain support by identifying the needs of their community and recognizing the importance of addressing those needs, business entrepreneurs develop charisma to connect with their clients. The individuals on the forefront of scientific advancement should also be encouraged to consider the value of knowing how to communicate with others outside of their field. Every day society becomes more adept to new technology showing a willingness to embrace change. However being socially adept is not a common ideology within the credentials of science. This is a dynamic I feel should be reevaluated. All specialized knowledge has the potential to be complemented by culture. Culture is what drives the need for science. Approaching individuals with the intent to make scientific work relatable may not only achieve understanding but also create a larger and more diverse group of people in lab coats.

Finding the value in making specialized fields interdisciplinary is accepting the fact that we live in an interdisciplinary world. It should be an understanding implemented in the curriculum of future professionals who are expected to solve future problems. Even the most contrived environments can't escape the potential issues that tend to arise such as public disapproval, inaccurate media representation, or lack of funding. These issues can impede on future discoveries made through research. Perhaps diversifying the minds of the professionals we depend on will ease on-lookers and encourage a broader understanding that can only be achieved through cultural competence. Having the support of the people is empowering to both communities in times of adversity.

When there is a blackout in the lab and researchers are scrambling to save lifetimes of data support should be found in various interdisciplinary relationships. Maintaining a relationship means a two-way exchange of insight. Such insight provides the tools to navigate the interdisciplinary world we share.

Michelle Mendez: Everyone is Included

The common stereotypes of scientists are people in white lab coats who are not very personable, however, every scientist has a relatable personal narrative and are very much belonging in the public as anyone else. Scientists and the public are not two opposing entities that constantly clash but rather one entity with the same fears, the same worries, and the same hopes in or outside science. The public is everyone and vice versa. It comprises of active citizens with diverse backgrounds ranging from different ethnicities, socio-economic backgrounds, and even different professions. When scientists "take off the lab coat", they belong with this public.

One example is that of Doug Melton, a scientist at Harvard University, who is learning about diabetes like any ordinary citizen with the intent of becoming informed. In the public, there are many who are diagnosed with type I diabetes who require a daily dose of insulin in order to survive, but this doesn't necessarily mean that everyone is up-to-date to the latest details. Before being labeled as just a scientist, Melton is also a father. Melton's two children were diagnosed with type I diabetes and like those who are directly or indirectly implicated, a great sense of worry and fear grows; Melton, like the public also does, invests time to learn about this form of diabetes to help his loved ones overcome this illness. Everyone has a different approach to take action whether it is seeing a specialist or researching online, but the main point is that being part of the public means taking initiative and in Melton's case, he works on beta cells from stems cells that might treat type I diabetes. This example illustrates him being labeled as a scientist does not imply that he also completely understands diabetes. Similar to the average person in the public, he too must learn about this condition because it may be something that is new that has now impacted his life and now must educate himself on this topic.

Another riveting example of a scientist fitting right in with the public is Doug Schmidt, a scientist at UC Berkeley, who was interviewed by our group. In Schmidt's case, it is the classical example of an individual being curious, determined, and creative—ultimately incorporating themselves into their work. In the interview, Schmidt mentions that he was intrigued by science at a young age in junior high as well as throughout high school because of science demonstrations and inspirational teachers. This shows that scientists are just an individual inspired to pursue their dream just as an artist is inspired by masterpieces or a politician inspired by a head figure—they are all inspired people of the public. Similar to many citizens, scientists are pursuing their hobby and are productive citizens alongside everyone else.

This list does not stop here. The main point is that when scientists "take of the lab coat," both literally and metaphorically, they are still human with the same glint in their eyes, and the same concerns of wellbeing for their family and their community. The lab coat is a façade but by no means meant to separate themselves from everyone. Everyone has a story and through these narratives, it makes us all more similar rather than dissimilar—we just have to uncover them.

Lucero Amaral: Scientific Perceptions

The word "scientists" evokes several, varying images and sentiments in individuals, which can be extremely positive, or reeking with negativity. This is due to the prevalent perception that scientists lack the normality that is so commonly seen in society, such as social abilities and morals. To further investigate this, I asked several people, with varying backgrounds and educational levels, to provide me with one word, or a few if needed, that they feel encapsulates scientists, or the scientific community, and the results were fairly fascinating.

The results showed a huge variety of answers and preconceptions about scientists and the world that they thrive in; the most common word was "intelligent", and its variations. Along with this, the second most prevalent were answers consisting of famous scientists, such as Bill Nye and Dr. Frankenstein, which imply that the media has a strong influence on how we perceive scientists. I considered these words to be in the moderate range, meaning that people who provided such words did not feel too strongly about scientists, and the work that they do, whereas other individuals provided words that gave a clear and precise view on their stance.

Those with a strong science background, or inclination, provided me with words that painted a positive picture of scientists, such as satisfaction, future and innovation. These words imply that scientists are doing work that is bettering the world, as well as held in high importance. Along with this, words such as magic and curiosity were provided, which paint science as being full of life, giving it a dose of humankind. These words showed that scientists are humans, like everyone else, following their passion and contributing, in the way that they feel most comfortable, to the general welfare of humanity.

On the other end of the spectrum were those that felt as if scientists lack morals or general sensitivity. The words that emphasized this were God and murder. These two words already have strong societal attitudes associated with them, but when used to describe scientists, are particularly negative. The word "God" implies that scientists are interrupting the natural process, an act that many feel is wrong. As for "murder" I asked the person who provided me with this word to further elaborate since it was not readily clear, and she said that this was in connection with animal testing. The implication here is insensitivity for all types of life.

As mentioned before, the majority of individuals provided fairly neutral and typical words such as empirical, factual and intelligent, words that show competence but not humanity. Since these words are typically heard, I wanted to unearth where these perceptions stemmed from. The answer would be obvious among my results, that being the media. This relationship is implied through the following answers obtained: Bill Nye, Dr. Frankenstein, The Big Bang Theory and Jimmy Neutron. The Big Bang Theory and Jimmy Neutron focus on individuals that are scientifically inclined, who are depicted as loners and unsociable, whereas Dr. Frankenstein is viewed as an overly ambitious scientists with no regard to life or consequences. These paint scientists as overly ambitious and having sacrificed their social life, implying that scientists are incapable of having a family or friends, outside of the scientific circle, simply because they chose to pursue science instead of any other field.

The current perceptions are prevalent because of popular media, therefore, in order to change the current perception, the media that propels the idea that scientists lack sensitivity and are basically living computers, needs to start showing that they to are humans, with fears and desires, just like everyone else. This is a difficult task since the driving force behind media is not necessarily truth, but what draws attention, and unfortunately, the perception that scientists are normal human beings, who are not trying to play God or defy nature, does not sell.

All Words:

Typical/Neutral: smart, assured, irrefutable, descriptive, concise, event-based, knowledge, accurate, unique language, interesting, intelligent, demanding, competitive, crazy hard, interrogative, evolution, empirical, experimenting, factual

Media Based: Jimmy Neutron, The Big Bang, Theory, Bill Nye, Dr. Frankenstein

Favorable: curiosity, magic, exploration, satisfaction, innovation, future, amazing, eternal (always something that needs to be answered – never ending),

Unfavorable: nerds, assholes, GOD, murder (with respect to animal testing), confusing, migraines, tired

Gurubala Kotta: How Can Scientists Take Off the Lab Coat and Why is it So Hard To Do So

Since no one is an expert in every field, scientists should not be seen as an elite and distant group, nor should scientists see themselves in that manner. As the name of our group implies, scientists should be able to "take off their lab coats" and respect those that do not wear the metaphorical lab coat. I believe that the metaphorical lab coat represents specialized knowledge scientists gain from years of immersion in school and research settings. It is important to note that having specialized knowledge should not be used as a means of determining "who is better". Essentially "taking off the lab coat" means respecting differing opinions and working together to break down hierarchies and stereotypes.

However, it is clear from our guest lectures that particular stereotypes have alienated scientists from those who don't wear the lab coat. Bridging that gap can have phenomenally positive effects thus I have chosen to examine what stereotypes scientists think are contributing to the gap. I will tie those findings into my paper and establish, from a scientist's perspective, why the gap exists and why it is so difficult to "Take Off the Lab Coat". Ultimately, my findings have drawn me to conclude that a scientist engaging in community outreach is an effective method of taking off the lab coat.

I became interested in this topic after Randy Schekman argued that the gap is mostly because of the "public's ignorance". His statement opened my eyes to the inherent biases of professors and researchers, and after Dr. Charis Thompson emphasized the important of not "accusing people of ignorance", I decided to delve into this topic. I will be analyzing hierarchies between scientists and the public, but mostly how best to mitigate those hierarchies. To do so, I interviewed a female neuroscience graduate student at Berkeley. I specifically chose to talk with a graduate student instead of a professor because I believe that at this stage in her educational training, she is most impressionable. She is being exposed to PIs, post-docs, professors, ethical dilemmas, and university culture. Thus she is in the midst of formulating particular attitudes and stereotypes of science that will stick with her throughout her careers. I asked her questions regarding her attitude towards ethics, the gap, and how best the gap can be bridged. The interview was incredibly insightful and I will discuss her responses and their implications.

Throughout the interview, I liked that she did not fixate on who is responsible for the gap; rather she delved into the importance of both parties being on the same page. Firstly, she said "how is the public going to educate itself about science without some inputs from scientists?" and stressed that scientists "definitely have a responsibility to communicate important scientific information for people who are eventually going to do things like vote on related policy". She highlighted that both parties play roles equally important roles—scientists have a responsibility to conduct research and communicate their specialized knowledge and the public has a responsibility to apply that knowledge to issues such as public policy.

Interestingly enough, the graduate student I interviewed pinpointed inaccuracy in science reporting in the media as the biggest cause for the gap. Earlier in the semester, we had a lecture from staff at the Berkeleyan, who said that when they write articles, they choose topics that are the most "eye-catching" and use "buzz words" to garner attention. The grad student I interviewed argued that most articles about science aimed towards a popular audience "are inaccurate at best, deliberately misleading with 'buzz words' at worst, and almost always have nothing in common with the actual published science". The inaccuracies and misleading information lead to conflicting reports and "ultimately the appearance that we don't know what we are doing, hence the mistrust in popular opinion". She also argued that nothing would do as much immediate good as "transparent and accurate communication of both basic science and cutting edge research to the general public by the media".

Her next few thoughts resonated with me the most; quite often, various media outlets insinuate particular stereotypes about scientists that make a scientific career look boring and unfulfilling. The grad student thought that the biggest misconception the public has towards scientists is that "we're all old white guys in lab coats". But she argued that it is easy to change that opinion—elementary schools tour NIH labs and do an exercise in which they draw what they think a scientist looks like before they go, and then again after spending the day with the researches there. The two images are like "night and day"—everyone starts out drawing the aforementioned "old nerd in a lab coat" and leave drawing someone "who looks a lot more like they do". As a parting thought on this subject, I believe that universities have an exceptionally critical role in community outreach; inviting kids to tour labs can have phenomenally positive ramifications. Disproving students' misconceptions of scientific research and life as a scientist at a young age can change their entire outlook of the field. Thus, universities should encourage each lab on campus to host elementary students for a day.

Lastly, I loved that she does not just "talk the talk"; she substantiates everything she believes in with actions. She engages in science outreach and her love for science proves that she is not just someone in a lab coat—she is a fellow human and part of public, not just a member of the scientific community. She travels to elementary schools and hosts "Mind and Brain Nights", where her and her group bring activities related to neuroscience; they try to get both parents and kids excited about how their brains work. Secondly, she volunteer teaches at San Quentin State Prison; she strongly believes that "nothing is more transformative than education" and argued, "as someone privileged enough to access a ton of it, I should do my best to spread it around".

Interviewing her was amazing, and I am grateful to work with someone with such a transformative view of the gap. As she explains, it can be difficult to "take off the lab coat" because of stereotypes and misleading information from the media, but it is definitely not impossible. Community outreach is a great way of spreading scientific knowledge and a huge thank you to her for being an inspirational scientist!

Sergio Reyes-Alejo: Taking off the Lab Coat

Scientists are often the subject of two prevalent attitudes in society: they are innovative and ingenious or that they are disconnected from the world and arrogant people with superiority complexes. Yet many people don't realize that scientists are people just like anyone else. This is highlighted by the fact that they are part of the public even within science itself; they only have expertise in one field of science and are a lay person in every other field, which is a huge realization considering the amount of fields that exist. It is this unrecognized truth that drives my project of studying the similarities in the paths of the "public" and scientists.

There exists many similarities in the life journey between both groups that are to this day still disconnected. I want to analyze these journeys throughout academics as well as life in general but with focus on communities of color, as they are greatly underrepresented in the sciences.

Our group goal is to see how different outlets have uplifted scientists on a pedestal and have created an ivory tower that acts as a divide but we want to debunk these assumptions. Because of this, I chose to focus on communities of color since their constant struggle to achieve in the sciences is strong and can often result in a strong reconnection with their culture, allowing the public to once again identify and connect with scientists. I want to see how science affects people of color in the rigor and strenuous nature that it entails throughout academia and compare this to people of color that do not pursue careers in the sciences or identify as experts in such.

In my experience, this reconnection with my culture occurred as I progressed through the science pre-requisites in my Cal career. The constant reminder that it would be an uphill battle and the lack of my people in my classes pushed me to strive towards my goal of becoming a scientist, all while not forgetting my community and relating the ideas my culture carries as well as my identity as a xicano.

Before coming to Cal, my connection to my roots, culture, and indigeneity were not as strong as they are today. As I took more and more science courses, I learned the hard way that the sciences were taught in a counter-intuitive manner. Testing our knowledge on the material was based off exams that would essentially pit the students against each other, but how can this be when science was a cooperative field? Dr. Matsui's lessons highlighted this very issue and his words resonated in my mind from his IDS 96 lectures. "Professional is personal." I always took these words in my thoughts. I contemplated what they meant to me. How could I take this concept to work for me and what did it mean about the professional I was to become?

This led me to really think about my identity. If professional really was personal then that would mean my positionality, my identity, and all the ideals that come with them were certainly to be taken into account when considering my science career, my stances on scientific issues, and ethical issues. Though I had met many other students who were against the way science was taught, it did not change the institutional placement of the curricula and its methodologies. This sense of isolation in science education beckoned me to search for a sense of community elsewhere. As a result, my identity was reaffirmed and restructured to encompass a strong bond to my cultural roots. My community was my people, my indigenous roots. My xicano identity served as a release from the strong institutional barriers set against people of color in the sciences and academia.

Sheelah Bearfoot: Taking Off the Lab Coat: The Mini Documentary

The "public", of which we are in some way or another a part of about many particular subjects, has varying degrees of contact with science. Because of this lack of exposure, mistrust and misinformation fill the gap between the public and scientists. People often question whether a scientists intentions for getting involved in a particular field of study are positive ones. My portion of the project focuses on bringing to light the different motivations people have in pursuing science as a career. All scientists have a moment, or a handful of experiences, that drew them into their work. All scientists have bad days where it's hard to keep going and easy to forget why they had a passion for this stuff in the first place. That story is what I want to draw out.

Throughout interviews and surveys, we found a treasure-trove of insight and a project that really took on a life of its own. We got pages of responses and almost an hour of interviews that we intend to compile into something bigger than a 5 minute video. These revealed differing interpretations about what "the public" thinks of science – everything ranging from "Intelligent" to "Future" to "Murder" to "Jimmy Neutron". Quotations from scientists about their work and what they think about public outreach: "As soon as you make a connection with a person, you can use what they know to help them understand it." Insight into their opinion on other scientists:

"My work is with people who often don't speak or write or read a European Language and they are often in remote areas of tropical countries where they speak an indigenous language that's passed on orally, so they don't have a written language. So by definition of many global organizations that would be a village that is full of illiterate people, okay. And I take issue with that. These people are actually very knowledgeable scientists on the local ecosystems in which they live." And even just listing hobbies – scientists play the flute, crochet, read, watch movies, and have fun at lab pumpkin-carving parties.

A lot of my own motivations behind this are from a discussion I had with a non-science major friend who told me "I don't really think of scientists as being curious. To a certain point I just see them as people doing their jobs." I really want to provide a narrative counterpoint to that remark, drawing off many of the experiences of scientists right here at Berkeley. I want people to see what a scientist is like when their lab coat is off. Because it really isn't that different from when it's on.

CHAPTER 5

How science situates itself in a larger world

One lens can still often result in two perceptions.

Chapter 5: Interpolation of Science

By Sabrina Rentschler

How does science situate itself in the larger scope of our everyday lives? Is there any way that we, the public, may be able to influence the rate at which discoveries are made? Has the advent of "pop sci" or popular science cheapened or misrepresented scientific concepts, or has it empowered the lay person by allowing once privileged information to reach the public? In this chapter students hope to provide a broad perspective of what science means to us, and how it has shaped our lives.

Science and technology serve as great indicators of a country's success and are major drivers of productivity and growth. This scientific growth builds upon itself as the result of these advancements and are used to fuel the next level of innovation. However, at one point does mankind become so immersed in its own technological power that we fail to realize our own flaws? To answer this question, student Simon Greenhill, brings into focus the origin of a child's passion for science, and the intersections between the economic growth and scientific literacy. It is in these analyses that notions of the binary nature of humanities and science begin to dissolve and we discover the fundamental intersectionality of STEM.

One ubiquitous example of the intersectionality of science is how it influences our daily environments; science does this as it informs the decision making processes of marketers as they sell products. Matt Kurata discloses his own experience in marketing consulting and remarks on the surprising amount of scientific insight that is used to generate sales. He shows us that marketers employ social psychology and neuropsychology in an effort to understand their consumers and compares the way marketers study their competitor's previous marketing campaigns to how scientists design experiments and form hypotheses.

The pervasiveness of science has no doubt effected life in positive ways, but it has also unintentionally created a form of class division in its wake. Although science finds itself associated with nearly every facet of our daily lives, not everyone can become a scientist. Unfortunately, this is often due to academic structures and financial limitations. Familial relationships, exposure to science concepts and access to educational resources arguably predicate a student's success far more than potential or ability. This chapter ultimately hopes that a better understanding of our combined relationship with science, will generate more equitable access and participation.

Part 1: Intersectionality of Science

By reflecting on our experiences, studies, and interests, we investigate how scientific discoveries have affected our everyday lives.

Daniel Russell Cheung:

Majoring in history has allowed me to see how much "science" and history intersect. Not only does each important scientific discovery have a history of similar discoveries leading up to it, but events in history have spurred scientists to discoveries. This can be seen in many points in history. Nuclear physics was a huge factor in ending the war in the pacific during World War II and was a major factor in the Cold War (this is just one example). Science and history have a mutual relationship.

Many events in history have acted as a catalyst toward discoveries and inventions in different fields of science. When the discoveries in the industrial revolution were combined with the growing antagonism between European countries and nationalism, the result was industrialized warfare in World War I. The resources available to these countries were nothing they had ever before seen in warfare. The creation Machine guns, poison gas, etc. which intersect with many fields in science were the result of events in history. Although this need resulted in the deaths of millions of people, the relationship between events in history and the necessity to fulfill these needs with science is evident.

Likewise, the discovery, creation, and advancement in different fields of science have lead to major events in history. The creation of the atomic bomb by Americans during World War II led to the end of the war on the Pacific and a necessary end to the advancement of the Soviet Union into Asia. The fact that the United States had the atomic bomb is important in setting up the stage for the cold war, but it is important to recognize its role in ending WWII. Obviously, the creation of the nuclear bomb was a great advancement in nuclear physics, specifically nuclear fission. The paradoxical relationship between history and science is evident.

This relationship is often ignored when looking at the humanities and STEM binary. The arbitrary binary places these two subjects at opposite ends of the spectrum, when we can see that they are indeed related to each other and our everyday lives.

Michael Ferrin:

Technological fields have been gaining popularity recently. As younger generations
become more technology minded they should also start learning more about science, because without science, technology is worthless. Without the laws of physics and the understanding of chemical conductors, and heat regulators, the word "computer" would still be synonymous with "mathematician". On the other hand, without the wonders of modern technology, we would have no biology labs, no vaccines, no magnetic imaging, and no way to reach out to the public about urgent scientific discoveries. The two topics feed off of one another, benefiting and gaining knowledge as they do. This is not a special case; no matter what the discipline, science literacy is essential.

Professor Laura Nader is an outspoken proponent of science, but shamelessly bashes the era of "techno twits". While attending a recent lecture given by Nader in the context of interdisciplinary science, I was surprised to hear her views on the relationship of science and technology. These are views she has been sharing for over 40 years: Massive technological development hurts. This is a fact largely ignored by economic planners, technicians, and political leaders. In planning drastic alterations in environment that uproot populations or make old adjustments impossible, they count the engineering costs but not the social costs. After all, they do not think of themselves as paying the latter (Nader 294)[1]

While there is no denying that human developments are slowly destroying the earth, many students, myself included, were rolling our eyes at the way she blamed technology for these wrongdoings, while keeping the good name of science intact. In reality it is becoming increasingly difficult to break the two fields apart. She used technology as a way to dissociate science from ethics, as if they have no relation. It was easy enough to buy into the argument that the tech industry works without regard for the environment, while science is purely for the good of humanity, but in reality science and technology share largely the same ethical concerns.

The tech world needs to embrace science, and prioritize science literacy more heavily within our own discipline, as they are the ones keeping us afloat. MATLAB and Fortran, two well-known programming languages, used in countless applications, were developed by scientists with the goal of simplifying complex calculations. Science makes technology practical. In turn, scientists must to learn to utilize technology in their daily lives. When it comes to communicating with the public for example, it has proven difficult to share important scientific discoveries. In a time when analog mediums are quickly becoming extinct, it seems that "Popsci" is the most efficient and effective method of informing the public about anything. This typically requires an online presence. Technology becomes the medium through which science is accessible. By recognising the intersectionality of science, countless doors are opened. Every field stands to benefit, in a cyclical nature, from embracing the scientific community. Not only is a basic level of scientific knowledge crucial for our health, but for our professional lives as well.

Appreciation for the subject within technological circles is arguably the most important, as without technology, science won't stand a chance of getting noticed by all the other disciplines.

Katherin Calero:

Many people have preconceived notions about science in public health. Some of these preconceived notions are that public health is a field only for those who want to become doctors and take the rigorous science classes, such as organic chemistry, biology, molecular cell biology etc. However, there is an alternative route that people are not aware of or don't understand. This alternate route is pre-health rather than pre-medicine. Through pre-health one is able to work with the community in various ways. For example through health policies, epidemiology, nutrition, community health services, etc. The general role of science in public health is to promote healthy habits. Though, science plays a different role for each field in pre-health. My goal is to work in the community health services and aid underrepresented communities that lack the knowledge of having access to different health resources.

As a pre-health major I am required to take a variety of courses in different fields, such as psychology, sociology, microbiology and biology. As a second year student the science courses I am required to take are intro courses, such as MCELLBI

55: Plagues and Pandemics and MCELLBI 61: Neurobiology: Brain, Mind, and Behavior. After taking MCB 61 and MCB 55, I began to reflect on the role science plays in the communities' everyday life, and it is important for me to be aware of this in order to be able to help them out. Many underrepresented communities lack the knowledge of issues, such as plagues, nutrition, factors affecting their brain and behavior. Due to the discoveries of many scientists, public health is able to promote the proper health facts and risks to the public. For example, discovering the outbreak of Ebola shows science role in public health.

Science in public health helps discover the causes to health issues and the solutions to improve these issues through proper medication or other factors. Through the help of science, public health is then able to make the community aware of these scientific discoveries and ensure that the public is well informed and has access to the proper resources. As a future public health scientist, my goal is to apply my personal experiences of a non-scientist and use these experiences to help the community which sometimes is forgotten doesn't have the same sources as other privileged people and thus are not able to understand what is going on in the health field.

Eden Marquez de Leon:

Scientific literacy has become a prerequisite in our increasingly scientific, technological society. Countless studies have demonstrated that our current educational and societal systems perpetuate educational inequalities, particularly in STEM (Science Technology Education and Mathematics.) This suggests that there is relation between education level and scientific literacy, putting entire groups of people at a disadvantage socially and economically. Through my own experiences, I will explore the intersection between science education and social class.

Sociologist Emile Durkheim believed that high quality public education should be available to everyone due to its greater functional importance in modern society. Despite the fact that there is a multitude of aspiring young scientists throughout the country, there also exists is also a barrier between those of low socio-economic status and the accessibility of science education itself. Children from poor communities experience schooling differently that those from more affluent backgrounds. When comparing my "low-income" high school in Oakland to other local schools, I find that educational systems do in fact help "reproduce preexisting inequalities". In my school for instance, AP classes were limited, there was a lack of individual attention from teachers and "learning [was] defined through rote memorization."

This diminished quality of education thus impedes the pursuit of higher education, let alone within STEM fields. It is then left to each individual to rise above the system, or fall through the cracks of society. My interest in Biology, for instance, did not stem from my high school experience. Instead it began when I had the opportunity to attend a STEM-focused summer program at Stanford University for a few years during high school. Unfortunately, not everybody gets opportunities such as I did.

Having a more heterogeneous group of people involved in STEM would also help ease the discourse surrounding scientific topics, ranging from climate change to vaccinations, since more intimate communication would be observed. Through my own experiences and reflections, I believe that improving nationwide science education (education in general) would not only encourage intergenerational social mobility but would also help develop scientific literacy as a part of the culture capital of more people.

Simon Greenhill:

Science is the foundation of the world as we know it. From high-rise buildings to cars to the computer I'm typing this on, everything we use comes from some innovation in science or technology. However, it takes absolutely no understanding of how a sophisticated piece of technology works to use it—it would be absurdly inefficient for everyone to have to design their own motherboards, machine their own auto parts or build their own houses. The proliferation of technology is generally a good thing, except that it does nothing to educate the public about science. In this world of constant discovery, we must strike a balance between advancing our knowledge and making sure the general public understands what that knowledge means.

In the 1960s, the National Academy of Sciences reported that "we're raising a new generation of Americans who are scientifically and technologically illiterate." [1] Despite the massive Cold War-era advances in rocket and military technology, the NAS deemed that American educational systems were not training enough scientists and engineers. Similar rhetoric is often heard today. President Obama has repeatedly pointed to increasing employment in science and engineering as a potential economic boon. Perhaps more important, though, is that the general public—people not researching or pursuing degrees in science—have a broad understanding of and appreciation for science. (Burke 34)

Although not everyone can become a scientist or an engineer, we should all be able to read an article about science and have sufficient background knowledge to understand it. Moreover, today's science labs and lecture halls lack diversity, in part because passion for science is often instilled by a child's parents and not by that child's school—as such, lack of diversity and inequality are perpetuated when passion for science is passed down within closed circles, only rarely allowing outsiders in.

In MCB 15, I've realized that there exists a profound disconnection between the researchers driving scientific fields forward and the rest of the us. Whereas researchers are incentivized to get published, win awards, and secure tenure, they might have more social impact if they split their time between research and education. Instead of trying to publish prolifically, maybe scientists should make sure people both inside and outside of their fields actually understand what they've been working on.

The conflict between researchers' need to publish and the social need for better science education was both implicitly and explicitly addressed in MCB 15. We heard explicitly from a number of IB or MCB Ph.D. students and postdocs, many of whom lamented, or at least mentioned, the constant pressure to publish papers, apply for funding, and move up in academic hierarchies. More implicitly, I noticed that the same pressures still seem present much higher up the ladder, even among tenured professors. I specifically remember one of our first lectures, in which Professor Bob Full could not stop mentioning Professor Randy Schekman's Nobel Prize. Even at their level—both men are full-fledged professors in their departments—it seemed that Full had an inferiority complex when Schekman was around. In reality, though, most people wouldn't really know what a Nobel Prize is, and even if they did, they probably wouldn't understand Schekman's work.

I don't mean to diminish the accomplishments of top-level researchers, but rather to question the balance between advancement and understanding in science. As science and technology continue to advance, there is a fundamental need for the general public to understand, at least on a base level, what's going on inside the lab and what that means for the outside world. Without that knowledge, we cannot question, challenge or guide scientific advancement, allowing an undiverse, achievement-oriented group to dictate the direction of scientific advancement. Science does not exist in a vacuum; it is fundamentally intersectional, impacting all parts of society.

It seems logical, then, that we should ensure that everyone is society have access to basic scientific knowledge. Through my experiences in college and particularly in MCB 15, I've come to the general conclusion that college is quite late to begin bridging the gap. By then, those who choose to go to college have often already established their interests and aspirations, making it difficult to go out on a limb and take science classes. The key, in my opinion, is to improve science education in elementary and middle schools. It's with basic knowledge, fun-filled lab activities, and science fairs that we can ignite young people's passion for science. In doing so, we can begin to bridge the many gaps in knowledge, opportunity and interest that exist between researchers and the public.

Matt Kurata:

As a business major, science seemed like a far away land that I would never venture to because I had no purpose to do so. However, after learning about the public's understanding of science and listening to numerous guest speakers, I not only learned that science must possess an important role no matter what your interests are, but also that science proves integral in almost every occupation and field of study.

Going off my own habits, I never paid attention to new scientific findings or sought after scientific knowledge because I felt as though it would be pointless since I did not want to pursue a career in science. However, I realize now that science can be of interest to everyone because it is integrated into almost every field of study. It may not be chemistry or biology, but it is nevertheless science and once the public realizes this, they will be much more inclined to pursue further information on scientific topics; I know that was the case with me. After learning that we are all scientists in our daily life, whether we are cooking, watching television, or exercising, I decided to explore science's relevance in things I am passionate about.

One of my specific interests and prospective career path is marketing. After gaining some experience in marketing consulting, I've noticed that psychology plays a crucial role in reaching the public. Every choice the designers and the consultants make—the color, the font, the medium, or the sound—has such an impact on the consumer's willingness to buy a product. All of the rhetoric involved with reaching the public is heavily rooted within social psychology (decision making, human's tendency toward holism) as well as neuropsychology (how the brain perceives images and sounds). Another key aspect of marketing is referencing other companies' marketing campaigns and techniques and using them as case studies and testing out certain techniques through focus groups, which proves analogous to conducting controlled experiments to test hypotheses.

Another passion of mine is entertainment, and science's role in movies particularly surprised me and piqued my interest. After listening to UC Berkeley integrative biology Professor Bob Full speak about his partnership with Pixar to create anatomically correct insects in the box office hit *A Bug's Life*. This led me to research more and find out that Disney's *Frozen* sought help from mathematicians and physicists to help them depict snow and ice in the most realistic way possible through formulas and other scientific facts.

Science is everywhere. Science is not just in biology books or in a chemistry lab. Once I realized that science was involved in movies and marketing, I couldn't wait to learn more about it. Once the public realizes that science is integral in their own passions and activities, I believe the so-called "gap" will be lessened.

Chloe Tsang:

"Hi, Welcome to the Berkeley Free Clinic. Thanks for coming out to the lottery tonight for dental services. I know that this process may be new for some of you, but while the ticket is being drawn, we would like to make maximize your time here by having a brief oral health discussion. Hopefully you can learn something new or share any insights you have."...."So can anyone explain what cavities are and how they form?"

"Holes in your teeth."

"Sugar?"

"Sugar and bacteria? They stay in your teeth so if you don't brush, they form holes."

"That's on the right track, but actually acid causes cavities! In almost everything you eat, there are sugars. Bacteria love these sugars, but when they eat them, they also produce an acid that gets onto your teeth and wears away the many layers, causing a hole."

"Now can anyone name a few things that could cause cavities?"

"Not brushing or flossing"... "Eating lots of sugary foods"

"Right! Anything else?" ... "Well there's actually many other things that could lead to cavities, such as the acidity of your saliva, common medications that cause dry mouth, smoking, drinking, using drugs, bottle feeding your child, and much more." ....

This personal account is one of many in which I've been sandwiched between a slice of buttery scientific knowledge and a slice of thin public understanding. Acting as the jam that fills in the gap, I've come to realize that it just takes a little smearing to get both sides to stick together. What I'm trying to say is that often times, what seems like elementary knowledge to scientists maybe in fact be completely foreign to non-scientists. What hours, years, and centuries worth of research and discovery found are in fact tremendously important, but doesn't change the lives of the public if only the details of the studies are thrown at the other side. On the other hand, remaining a vanilla white piece of bread doesn't help. The public needs to be active, curious, and engaged in embracing the buttery knowledge. And how does this come about? It's all on the jam. The jam is where the intersectionality of science and other fields need to smoosh together. This is where there is no clear divide. This is where science is receptive to learning from the public and the public is open to receiving from science.

Particularly in health care, there is a lot of room for the jam to be spread. As biotechnological companies, pharmaceuticals, clinical hospitals, and many large corporations develop innovative ways to improve health care, there needs to be a way to funnel all those discoveries effectively and efficiently to the people who need it. For the public, there needs to be avenues in which they can turn to for streamlined, accurate, and engaging information. In this intersection, presentation is key. How well science is able to relate to people equals how willing people are to learn and making any health changes. For instance, the recent Sugar Tax passed in a few states makes it mandatory to charge a tax on sugary drinks. From the scientific perspective, this will drastically cut down sugar intake which has been widely correlated with obesity, diabetes, and other cardiovascular diseases. Not only will the sugar tax hypothetically improve health, economically it will bring in revenue. However, from a public perspective, a limited understanding of scientific reasoning could lead to frustration as some may argue that it's an infringement on rights. As a result, the intersectionality of science and other fields is vast. However, it could be a space for conflict or one that fosters open communication. The important piece is that there needs to be accessible avenues for feedback and discussion between science and public health. These could be in the form of media, education programs, scientific academies open to the public, demonstrations, public talks, or much more. So let's hope that as our scientific knowledge expands, so does our ability to find ways to engage each other.

Part 2: Science at the Interface

The goal is to improve various applications of science and health in daily life by addressing key issues of communication and interaction with the public.

Jessica Yescas:

The access to healthcare nowadays with the implementation of online enrollment should be all-accessible to those who are in need of this care, minorities in specific. However, this is not the truth in many cases because of the lack of access to these resources. Communication between those helping with enrollment and those attempting to enroll is not working as well as it should be. The fact that language, geographical location, and cultural awareness continue to be barriers with the healthcare enrollment system is why these key issues need to be addressed. This thoroughly relates to my group's overall goal of attempting to improve the applications of science and health in everyday life by interacting with the public. We will have to analyze what approaches to this issue are working and which ones are not in order to bridge this health care gap. My topic also relates to the overall chapter because it is necessary to have a good background in the sciences in order to be able to use them in our everyday lives. Science affects all people, either directly or indirectly, what we have to figure out as a group is how to have the public have the information and access to utilize all that the scientific world offers them. I am most interested in this topic about the interface, because I feel it is the most inclusive and relevant to all aspects of the sciences. There is much room for debate and research in this topic, which seems interesting to me. Of course, being a public health major is part of my motivation to want to work through this topic. So many resources are out there nowadays concerning this subject, online resources specifically. Articles, books, and researchers all have extensive research on a variety of the topics that this category of "Science at the Interface" entails. I am currently doing a research paper on this topic in particular, and have already found various online sources about the issues surrounding healthcare. Because of these issues, I hope to offer up solutions that my group and I come up with.

The issue of money comes along with a people's access to the Internet, due to the cost of Internet implementation in a household and the socioeconomic status of the people. Many minorities tend to be low-income as described in the 2012 U.S. Census, which illustrates that Hispanics and Blacks were at an average $20,000 below the average household income in Whites. Minorities do not have the money to spare for the access to Internet, meaning that they will most likely have no manner in which to access this online healthcare. A study done by several healthcare professionals at the American Medical Informatics Association titled "Household Computer and Internet Access: The Digital Divide in a Pediatric Clinic Population" indicated, "Internet access was found in 86.3% of households earning at least $75,000 annually, compared to only 12.7% of households earning less than $15,000". Additionally, in people with a household income of less than $30,000 it was demonstrated that they had about a 40% search for online information. The low-income households then prove to have a lower overall search for health topics because of the correlation to no access to Internet. This fact associates with a study done by the Agency for Healthcare Research and Quality, which found that "people in poor households had worse access to care than people in high-income households on all access measures. Blacks had worse access to care than Whites for about half of access measures. Hispanics had worse access to care than Whites for two-thirds of access measures" (7). These studies have shown that those groups of people with the least access to Internet are in fact minorities. The association between the low income of minorities and their lack of health insurance demonstrates that there continues to be a healthcare disparity; this disparity is in those who live right above and below the poverty line, an average of 26% of Hispanics and blacks lived in poverty since 2012 compared to a 10% average in Whites, and who lack Internet access (U.S. Census 2012).

A solution to close the disparity gap between Whites and minorities in healthcare would be to hold events for healthcare enrollment in their communities in order to bring the Internet to minorities instead of having them go through the costly process of searching for it for themselves. Of course, the most ideal situation for minorities would be to officially have them all have Internet access, but these events would be just as helpful. The enrollment events would not only help close the Internet access disparity or the economic gap, but also the education gap in Hispanics and African Americans, as the events entail assistance provided to them during the enrollment process. These enrollment events would also be free of cost and be run through volunteers and charities, allowing for minorities who are also low-income to gain access to these resources. With the solution provided, we can only wait to see the results years from now, when a new generation of minorities will potentially demonstrate if there is improvement in health or not due to the healthcare provided by The Affordable Care Act. If the solution provided is implemented and has positive results (i.e. larger amounts of Hispanics and African Americans enroll in these healthcare programs and are able to receive quality healthcare), the health of these people will begin to climb, potentially projected to being in line with the average health of non-minorities.

Osman Shokoor:

One of the largest changes in recent U.S. policy has been healthcare reform and as people hear buzz words such as "ObamaCare" and "Affordable Care Act", many do not understand the changes that have been made and the measures need to be taken. Often, the only time a person or family is exposed to healthcare is at the doctors or through the news. The public (clinical patients) understanding of healthcare policy is crucial for a better healthcare system and for a healthier population. Although measures have been taken to communicate healthcare changes, barriers to understanding still exist that limit how one interacts with the healthcare system, ultimately affecting health decisions. For example, if a family does not learn about the different health insurance plans, they may choose an unideal plan that can hurt them more than help them. If a family does not understand the requirement of receiving health insurance, the family may be subjected to a penalty that can harm them economically, and indirectly affect their health.

Doctor-patient interactions can be the key bridge to the gap in the public knowledge of healthcare. Doctors should inform the family about the economic repercussions of a procedure and surgery, and should ask the family if they have any questions regarding their healthcare or health insurance. Those without health insurance plans should be consulted with to analyze what the best option is for the family. Health care practitioners should be trained to recognize social determinants of health and to advocate for healthier lives for their patients. Physical or clinical health isn't the only factor in overall well-being as economic and social health greatly shape an individual's or family's lifestyle and health outcomes. I hope to continue to analyze ways in which people can learn about their healthcare plans and US health policy conveniently and to catalyze healthcare by connecting people to resources in the community to do just that.

A large exposure to science comes from an individual's or family's experience at the hospital, clinic or doctor's office. Health is an essential aspect of everyone's livelihood and proper communication is helpful in bringing proper healthcare and in advocating for positive health outcomes. Healthcare policy, health insurance, and health decisions and diagnoses of health conditions can be a nightmare to properly comprehend and many barriers arise that can lead to limited knowledge and uninformed decisions. For example, lack of English fluency acts as barrier to proper communication. Thus, interpreter services should be an essential aspect of proper healthcare.

Clinic integration of social determinants of health is crucial in advocating for proper health outcomes. A new standard of healthcare can be put into place where health care professional can inform patients about lifestyle choices and environmental conditions that can affect health. For example, I volunteer as an Advocate at Kaiser Permanente Hospital in Richmond with Health Leads, a non profit organization that seeks to catalyze healthcare and connect patients with resources in the community to promote healthier lives. This new model of healthcare, in which social determinants of health are directly addressed in a clinical setting allows for proper dialogue and engagement between patient and provider in working towards physical, mental, and social well-being. Healthcare providers who identify social barriers to proper health of a patient can refer patients or families of patients to community resources or social work and patients can utilize these resources to ultimately gain positive health outcomes for themselves or their family and may indirectly lead to more economic productivity and an overall healthier population in the U.S. Continuously rethinking the status quo among systems of healthcare is crucial for creating positive change that will increase communication between health professionals and patients.

Jeffrey Zhang: The Power and Influence of Popular Science

Almost no one likes to receive confusing information that they cannot understand. Science is a prime example of this. It is notorious for being extremely difficult, both to learn about in class and even to comprehend the concepts. There is no way that I would have been interested in the sciences if I had first been exposed to the wave functions common in physics and forced to immediately decipher them (I can't even figure them out now.) Of course, science is a difficult discipline to fully learn, but that does not mean it must be, at first. This is the value of popular science media, commonly abbreviated as popsci. Unlike a scientific text or research paper, which is designed to be concise and crammed with specific information, popsci articles only give the conclusions, theories, and results of various experiments, which is much easier for the average person to digest. In addition, these articles focus on a more creative tone writing, making them more interesting and engaging to read as well. Not saying that they have to be articles either; popsci can be in any form of digestible media. Popsci can be for all ages, but I find the best popsci to be geared towards children.

I myself was, at a young age, driven to pursue the sciences by my exposure to many different popsci resources, of which included a vast collection of popsci books and DVDs (all of which were incredible to read and watch), multiple trips to the museum and BrainPOP (this was the stuff back in the days of elementary school). This gradually built up my interest in science, which has allowed me to somewhat understand and have fun learning all of the material required for a chemistry major. This relates to our group goal as we are trying to improve science in daily life by resolving communication issues, which would be considerably easier if the public was eager to learn science as I am. In order to do this, we must expose kids to more science! Imagine how excited a small child will be when they see a dinosaur exhibit, or a shark documentary, or even a Bill Nye chemistry show for the first time? They will be excited and ask their parents question after question on what it means, who will in turn buy many books which simply and entertainingly explain the science, and as a result get the kid more excited in science. The parents will in turn be more interested in science as well, given their kids interest, and the whole family will be more willing to learn this strange but fun phenomenon that is "science!" They will consequently be more willing to listen to the scientific community; the interface between the public and the scientists will be a little easier to cross.

My specific role in "Science at the Interface: The Interpolation of Science" is in spreading the interest of science via popsci by sharing my experiences with it. This will increase sciences role in the world and spread knowledge of it to more people. I know that it worked for me. So why shouldn't it work for others?

Vin Y. Lay: Science Education in America

We know that science is important because it exists everywhere in today's world whether it be grocery shopping, cooking, or recycling to online networking, curing disease or influencing public policy. However as the pace of scientific research accelerates, the remaining public has to increasingly grapple with matters within science. This creates more opportunities for misconceptions, for example, *science is just a collection of facts and is complete, science is so complicated that only those who work in the field of science can understand its nature*. It is important that the public understand what science can truly be and how public policy can be informed by scientists. It is what allows us to understand the causes and effects of conditions in human health and informs us of ways to stay healthy because we know that "smoking is a cause of lung cancer; obesity is a cause of diabetes; alcohol and drug use by pregnant women are a cause of brain damage to their unborn children" and so on. It is also through science that policymakers advocate or oppose certain medical research or laws that revolve around these issues.

Science has been a part of the American school curriculum for a long time, but is something students find tedious with details, and that they tend to forget the materials after exams. Even the basic curriculum taught in secondary school often consists of lectures on taxonomy and various facts, along with the excruciating memorization of strange terms. This may be another reason why there are so many misconceptions that form as students become older. I believe that in order to improve the way the public views science, the standard school curriculum must be changed.

Rather than memorizing the names of different parts of an organism, emphasizing the "difference between data and speculation, how to frame a question, and how to approach a problem critically and skeptically" would create a more applicable agenda for students[1]. I interviewed a UC Berkeley student, Betty Ha, who was also her high school valedictorian. Ha said that she struggled with courses in college even with most of her AP science courses being college level. She added that her AP courses only required her to memorize the material rather than learn critical thinking. Ha is currently a third year undergraduate student and working in a molecular cell biology lab. She mentions that courses at Cal have taught her to think critically about the "real" techniques and knowledge concepts that she needs in her lab, as well as for her summer research internship. These approaches allow students to be equipped to bring scientific thinking to matters and issues that affect their daily lives.

Another student, Seena Tabibi, graduated from UC Berkeley in the spring of 2015 and is attending Southern Illinois Medical School in the fall of 2015. I met Seena when he was a chemistry tutor for a class I was taking. Tabibi shared that because he had "a father who is a physician and a mother who is a pathologist, I was able to get a lot of advice for life, academic decisions and help with science assignments from my parents during my undergraduate career." However he admitted that most of his science assignments in college were memorization based, like Biology 1A. Therefore, no matter how much advice he got from his parents, at the end of the day he had only himself he could rely on.

To see the differences in science education between a two-year vs. a four-year college, I interviewed my friend from high school, Rebekah Quenta, who goes to Diablo Valley College (DVC). Rebekah is a pre-dental student who is planning on transferring to a four year college after finishing her general education courses at the community college. She said that in her chemistry and physics courses the professors encouraged students to use critical thinking applied learning styles because of the problem solving required in the curriculum. Still, in her biology courses, professors continued to use PowerPoint to teach which tend to be memorization based methods. Quenta pointed out right away, that biology must be learned through applying knowledge in a laboratory and in real life. This is important because unlike many four-year universities, professors at DVC do not emphasize research aspects of science. Quenta suggested that community colleges "should have students doing long term projects by using the knowledge that they gain in the classroom as a mandatory curriculum," because it allows them to better understand the material and be creative. Nevertheless, one thing she continues struggling with going to DVC is not only applying the material to real life but simply enrolling in courses is difficult. "There are only about thirty people in each class, which is good because I get to build good relationships and get more help from professors, but it takes a whole year to get into each course." The initial reason why she chose to go to a community college rather than a four-year college is to save time and money, but in some ways community college neglects a main aspect of science education-- accessibility. Most of the students who are in community college are either planning to transfer to a four-year university or go directly into the workforce. If real life application/research is not emphasized, it may already put many students at a disadvantage. Additionally, without accessibility for students, it may take them longer to finish college or delay their entrance into the workforce. This would ultimately defeat the original purpose of saving time and money by going to a community college.

The aim of this paper was to descriptively analyze different students' experiences on American science education. In doing so I hope to bring awareness so that the public may realize the many problems with today's science education and why many people have a negative perception of science. According to the students I interviewed, it would seem that the government could afford to improve accessibility for the students, as well as adjust the science curriculum. In doing so students may truly use what they learned in the classroom to influence life decisions such as public policy in science.

Pia Choi: Issues with Online PopSci Resources

The members of the general public who are not very educated about research and science have a tendency to be too trusting of claims made in popsci resources. This is a problem due to the fact that many people turn to popsci articles on the Internet to learn more about how they may improve their diet and in turn lead healthy lives. I will use an article titled "The Detox Prescription: One Doctor's 8 Reasons to Juice Cleanse" by Suzanne Hall in order to show the way in which people interpret and react to a popsci resource about diet.

The article talks about the benefits of doing the juice cleanse, which essentially involves cleansing the body of toxins by only drinking juices for 3 days in order to improve one's health. The article states that a doctor named Woodson Merrell wrote a book that promotes and explains how to juice cleanse. The article claims that by doing a juice cleanse that one will give their digestive system a break, have abundant energy, reduce inflammation, reduce cravings, lose weight in a healthy way, rehydrate, reduce chronic ailments, and improve skin tone.

Out of a total of 26 comments, as of April 27th 2015, only 1 person left a comment questioning the claims of the article. Most of the comments talked about trying out the juice cleanse and having success with the juice cleanse, success in this case meant that the commenter felt "good" and "healthier" after doing the cleanse. It was astounding to see how only one person questioned the claims and indicated to a potential negative to juice cleansing. The potential negative that the commenter mentioned was about how extracting just the juice from fruits and vegetables may mean that the many nutrients in the leftover solids are being tossed out.

Other than the potential waste of nutrients there were many other dubious things about the article that no one seemed to notice. For one, the article seemed to put a lot of emphasis on the credibility of the author based on the single fact he is a doctor. A doctor has no credibility talking about diet, much less releasing a book on one considering the fact that most doctors are not dietitians and they do not receive the education required to make them experts on the subject. Secondly, the article did not mention any negatives to juice cleansing, which include the possible side effects from the high amount of concentrated sugar in juices. Lastly, the article did not site or directly quote any research to back its claims. It mentioned that "Multiple studies conducted in Scandinavia in the 1990s proved irrefutably that juice cleanses followed by a vegan diet could powerfully reduce inflammation in the body", however it does not go into any specifics of the studies.

My overall purpose is not to discourage, accuse or slander this specific popsci article, or any popsci article for that matter. In fact, my purpose is to expose how members of the general public have a tendency to trust and endorse scientific claims about diet and health that they discover in popsci resources, especially easily accessible online resources, and incorporate them into their lives without once thinking of the debatable and uncertain aspects of the resource. Overall, reflecting on my group's goal to improve various applications of science and health in daily life by addressing key issues of communication and interaction with the public, I hope for people to be aware that there are online popsci resources that make claims without definite proof, to be cautious when choosing to believe or not believe these claims and if possible to do further research on the topic and more importantly on the claims made by the popsci resource in order to properly integrate authentic and dependable information about diet and health into their daily lives.

Anonymous: Improving Doctor-Patient Communication

Doctor-patient interaction is flawed in so many ways due to miscommunication. Doctors often withhold information from patients who are less educated, which leads to the patient's mistrust and failure in complying with treatments. Michael Balint's theory of the "apostolic function" which incorporates the doctor's belief of how patients ought to behave when ill, is highly evident among domineering doctors. This interaction devalues patients, making their opinions less important than what the doctor asserts about their illness. Such flawed system causes an ongoing clash which leads to a "vying for dominance by each participant," akin to a micro political system in which the physician uses medical knowledge to maintain dominance, while the patient continues to suspect the veracity of their physicians (Anderson). Doctors are also prone to making stereotypical diagnoses, without understanding the fuller context of each of the unique cases that a patient brings. Most often doctors underestimate the amount of knowledge patients have about their illnesses.

Effective communication is therefore vital for successful doctor-patient interactions. According to an instructional course called "Difficult Conversations in Orthopedics," team members discussed numerous techniques and methods to improve conversations. One is called the 4 Es model-engagement, empathy, education, and enlistment, which allows patients to be treated mentally and emotionally, not just physically (Leahy). Engaging with the patient using eye contact or an appropriate touch on the arm builds trust and a bond of sincerity. Empathy, different from sympathy, acknowledges the underlying emotions of fear and anxiety of a physical illness. Education or teaching a patient in their level of understanding is helpful, while enlistment or tailoring a treatment plan that best fits a disease is important. Finding a middle ground is important to developing a third culture for both the physician and patient to communicate comfortably and effectively.

In an effort to create a third culture between the patients, doctors should understand the greater context of where their patients come from by taking into consideration the cultural, economic, and psychosocial backgrounds. Social problems such as violence, drug use, depression, teenage pregnancy, and poor educational performance prevalent among the lower social class cause insecurities and stress that affect the ability to withstand disease (Wilkinson). As health is powerfully determined by social factors, it is vital that doctors address these issues and converse at a more colloquial level for a balanced interaction. Informing patients about social and educational services helps create a nurturing relationship between doctor and patient and allows patients to apply healthy habits to their lives more effectively.

A group in the Bay Area called the Highland Health Advocates addresses socioeconomic problems such as lack of housing and food to determine the root cause of illnesses. When I first applied for this program, it reminded me of MCB 15 and how it has taught me to become a better communicator through the numerous discussions we had. I thought that becoming a health advocate would allow me to develop interpersonal interactions that would serve to bridge the gap for those who lack the resources and knowledge. When I officially volunteer in the fall, I plan to use the communication skills I have learned from taking this class in order to become an effective communicator. Knowing that people come from a rich and diverse array of cultures and religions, I plan to do my best to tailor each patient in the best way possible. As an aspiring doctor, I hope to continuously improve my communication skills in order to treat patients with empathy and care.

CHAPTER 6

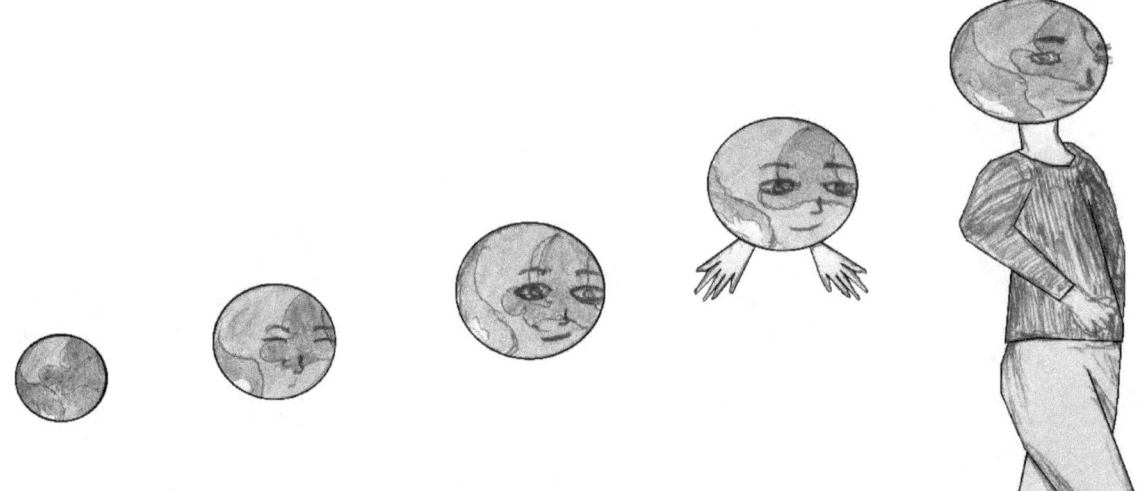

Class Dismissed! Now What?

An ever-changing discipline makes critical thinking a valuable tool, improving science education as well as gaining public trust and support.

Chapter 6: "Class dismissed!" What now?

By Danny Lee

Edited by Julie Mendoza

The developments in science are ceaseless, making it an ever-changing discipline. The rapid nature of these changes make it a difficult subject to translate into classrooms. A way to better prepare students is by providing proper tools. Critical thinking skills are powerful mechanisms when faced with unforeseen currents of change. The hope for an ethical and well-informed populace in the future relies on how we equip students today. Adjusting the scientific perspective of teaching by finding engagement through conversation and debate is another approach to actively shaping a matriculation that facilitates long-term improvements. Methods that encourage a two-way current of information allow scientific thinking to become applicable after class is dismissed where there is still plenty to be explored.

Education is also a reflection of access to resources. Classroom demonstrations that are comparable to laboratory practices, for instance, require special materials and a certain amount of modern equipment, which is not always readily available for those who are eager to learn. Students that have a stronger understanding also have better access to resources that provide exposure to authentic materials. In addition to stimulating the imaginations curious about science, contact with laboratory equipment give the student's interests worth and adequately prepares them for the pursuit of furthering their scientific experiences. The pedagogy of science reveals a growing disparity of knowledge within a system in which everyone begins with a clean slate and the ability to do well. Current inequalities continue to create an imbalance that only exacerbates as students advance. To combat this issue, we suggest that the pedagogy of science be flexible to various ways of learning and more available so that students become more deeply rooted in their confidence to learn.

In this section, we focus on the current K-12 systems in urban and underserved academic environments. Through discussion students recognize that there is a need for improvement, leading us to evaluate the possible benefits of modifying traditional methods. All reforms require total participation from everybody in order to be successful, as discussed in the section "Science and Pedagogy."

Positive reform of science within public education is dependent on the support and trust of the people. There are times when advancements produced within science seem to move quicker than we have time to evaluate their possible impacts. Genome editing and other discoveries, which may have never been pursued in the past, force society to accept what might have been traditionally deemed immoral. The response of the public outside of the scientific community remains an important factor in maintaining a balance and healthy tension between those who create and those who consume. The willingness for the scientific community to remain transparent reassures the public that all implemented changes are still in the best interest of the people. It reminds fast thinking individuals in the field that, above all, the nature of their work should remain safe, its purpose noble, and unintended consequences rigorously considered.

In "Future of Science and Change" we continue asking how to proceed ethically. The "case" of focus is genome editing as well as the controversy of designer babies. The thoughts and opinions of students are accompanied by their willingness to create a platform for conversation that takes into account the responsibilities of working in the field.

As science develops so must society in terms of staying aware of relevant ways to evaluate the policies and expectations for this discipline. It becomes necessary to revitalize discussions that hold change accountable for its wider implications without stifling it with socio-historical inertia. Providing a well-resourced and flexible education can be a novel way of setting a stronger standard for the future of ethics, knowledge, and change. This chapter explores specific sustainable models to make this happen.

Part 1: Science and Pedagogy

Understanding the pedagogy of science, how it affects perceptions of science, and how it should be taught to grade school students

Danny Lee: A Look at Challenging the Status Quo in Underserved K-12 Schools

Disparities in education reveal not only the social inequalities the public faces, but the challenges faced with trying to bring change to the current system. The pedagogy of science, for example, is rife with issues such as the curving grade system and memorization vs. application. As an undergrad student taking courses such as general biology myself, I would like to see change and advocate for students. Learning general biology may be interesting but the classes themselves do not encourage that. It would be great if there was some collaborative environment that encourages this type of learning. Students should be able to receive and use a "scientific", or different perspective, not just gain interest. Reform of the pedagogical system is needed to challenge not only students, but also educators and the curriculum so everyone can benefit from this change. The group's initial objective was to just focus on education on the K-12 level and find suggested methods of instruction. However, after talking with Professor Matsui, we decided to be more specific and identify teachers in underserved, low income, urban public schools as our audience because students in those communities need the most support with how our current system of education works. We chose to make our infographic catered towards teachers yet the audio would help anyone who wished to view it. This relates to the overall group's goal in investigating how the education system affects how science is taught and how it could be changed to provide a more accountable, sustainable ideal system.

Our meeting with Professor Matsui led us to rethink many of our previous ideas and approaches towards this project. He didn't provide answers but many thought provoking questions. How would this proposed model be sustainable and accountable? The point he brought up about how the California Academy of Sciences wasn't willing to change its educational methods to cater towards kids from the Mission District in San Francisco versus the Pacific Heights area visiting struck a chord with me, as I have visited the Cal Academy of Sciences twice as a kid. I remember going there for fun and not really "learning". We thought about various approaches and solutions, such as how feedback could be provided on the proposed model. We did not want to blame teachers, as they are passionate about helping students but how the system is set up so that they are limited in what they can do. Teachers are already underpaid and overworked so we didn't want to put another burden on them. The proposed model should encourage them to join the team and provide them with resources. We wanted to tweak how the education system functions so everyone's (parents, administration etc) opinions would be considered. Through discussion, we identified the current system, the ideal system, and how the ideal system could be sustainable.

I worked on the current system, and learned a lot from how the current system affects students in urban areas. Jeff Duncan Andrade, who started his own school in Oakland was really inspiring to me as he nurtured "roses growing out of concrete" instead of negatively viewing them. He brought up Maslow's hierarchy of needs and how these urban students don't have the basic comforts necessities to do well in school. Students need the initiative to learn but they first must be provided the resources to be motivated. Providing the resources would be difficult (sustainability) but we considered sciences "kits" and partnerships with NGOs and other organizations. A review board with the stakeholders in the new system would make sure implementation would be smooth or changed. The culture of compliance would be changed to encourage curiosity and diversity in many ways would be valued. Our infographic and audio will allow us to show how science education, not standards, can be improved to cater to a diverse group of the "public", from parents, teachers, and students.

For more information please see:

http://mcbfifteen.tumblr.com/post/118681426390/mcb15-the-pedagogy-of-science

Works Cited:

Duncan-Andrade, Jeff. "Growing Roses in Concrete" TEDxTalks. 11 Nov. 2007. Web. 8 May 2015.

"More Than 40% of Low-Income Schools Don't Get a Fair Share of State and Local Funds, Department of Education Research Finds." More Than 40% of Low-Income Schools Don't Get a Fair Share of State and Local Funds, Department of Education Research Finds. U.S. Department

of Education, 11 Nov. 2011. Web. 8 May 2015.

<http://www.ed.gov/news/press-releases/more-40-low-income-schools-dont-get-fair-share-state-and-local-funds-department-education-research-finds>.

Mohamed Taleb: Educacion Vs. Schooling

I define pedagogy as what you teach (content), how you teach it (creative approaches) and why you teach it (your passion). Looking at science from a critical, pedagogical point of view specifically K-12, most educators are doing a horrible job motivating, influencing and shaping the new generation to be scientists. There are numerous reasons of why educators are doing a bad job. The first is teachers, who teach in urban communities, are unable to connect to student experiences. They do not comprehend the daily oppressions surrounding low-income student's lives like poverty, violence, drugs or whatever it may be. Next are teachers who lack the abilities to engage their students in the sciences by failing to show the applications of the content into their lives and believing that their students can become the next Albert Einstein.

Nowadays students do not realize the importance of their education and the trajectory it will have on the quality of their life. This is a result of multiple factors at place but my main focus is the pedagogy of science teachers. Based on four years of research on the streets of Oakland (Doc your Block), teachers in different grade levels (Build Your Own Teacher), my school (Build Your Own School) and my community (Build Your Own Community) where I presented at the American Educational Research Association (AERA), I have found that teachers are not educating students like their lives depend on it. These are the young minds in which educators are shaping in order to become caring, contributing citizens and all these educators do is school our youth just for their paycheck. Gloria Anzaldua distinguishes between *educacion* and *schooling*. Educacion is the quality education that is culturally relevant and impactful in student's lives. Schooling is the institutionalized brainwashing of culturally biased information that de-motivates students from attaining their hopes and inspirations.

The successful educators, who practice educacion methods, that I have encountered are those who have established an authentic relationship with me in which they have become my mentors who have steered me in the right direction. One of the greatest educators and the most memorable mentors I've had is Jeff Duncan-

Andrade. Jeff has volunteered to be my English teacher for four years because his teaching method was to stay with a cohort of 30 students throughout high school while being a professor at the San Francisco State University. Jeff talks about three different types of educators: the *Gangstas* who are teachers who school and do not care about students, the *Wankstas* who provide surface level care for students that is coated with false hope and the *Ridas* who are willing to put in the time and effort to create curriculum that enhances student engagement and critical thinking. Going through the K-12 school system specifically in the science fields, I have come across Gangsta teachers who could care less for students. The most common teachers are the Wankstas who attempt to educate you but end up schooling you. They are the ones that are prevalent in the schools I attended where they seemed to really care about you being successful but do not create an environment where that is possible. The Ridas, like Jeff, come every so often as they are the few educators who teach like there is no other day.

What made Jeff a rida is his tenacity to revolutionize the education system. He has continuously persevered to bridge the gaps by de-constructing failing institutions that are foundational unstable and supported by pillars of inequalities. These are the type of educators who will motivate the next generation to become future scientists and these educators are the ones who need to be in the classrooms applying their pedagogy into the science field. From my research, these are the educators who succeed in influencing the new generation. As I reflect on science and pedagogy, we need educators, like Jeff, to motivate youth of color to revolutionize the STEM fields so these students can come back to their communities and continue the cycle of scientific pedagogy.

Victor Vargas: "When Will I Even Use This?"

When talking about her mentor Anne Mansfield Sullivan, Helen Keller said, "It was my teacher's genius, her quick sympathy, her loving tact which made the first years of my education so beautiful. It was because she seized the right moment to impart knowledge that made it so pleasant and acceptable to me."

The way that students are taught in classrooms today deals with an impracticality, which is the absence of applications and coherence to what they're being taught. In order to better understand this impracticality, we must keep in mind that teaching assistants, teachers, and professors are paid for one thing: to teach students their subjects; they aren't paid to, nor is it ever an objective, to get their students to love, become passionate, or to continue on with these subjects. If we look in the classroom setting, courses aren't taught with the objective of interesting students in the long run; therefore, teaching methods often neglect teaching in an insightful and coherent manner. Here's how it usually goes down: students are taught subject material in a very direct "one size fits all" way, students are then required to accept what is taught at face value, students then have to study for exams, taking in a vast amount of information that isn't very structured to begin with and then forced to throw it all back up on exam day. This method works for very few students; participation and enthusiasm is more common in the classroom if students are interested in what they're learning. If students were in fact interested then we could have a situation where most, if not all, students would wake up every day eager to go to school to continue their studies. This situation may not be as far-fetched as you think. A simple and efficient way to get students interested would be to show them how subjects being taught have relevance to their everyday lives. This would be done by relating a subject to something that can be easily identified by these students; an example would be the sport of Bowling, relating how the mechanisms of pinsetters deal with applications of Physics. As well as revealing and depicting the possibilities and capabilities that exist within certain fields.

Amongst many other subjects, I've realized that Mathematics in particular is taught in such a way that neglects its applications. I can recall Geometry and basic Algebra being taught this way when I was a high school student. For example, we'd learn how to solve and set up equations, learn to utilize formulas that were never derived for us and then expected to regurgitate all this uncertainty on an exam. When I was a high school student, I often felt that Mathematics was very linear, as did others, involving a simple "plug and chug" method of doing things. Now as a student in the University setting majoring in Mathematics, I've realized that Mathematics courses are far more than simply "plugging and chugging" as they require me to think more abstractly. Also, in order to be successful, one must have strong problem solving skills as they are tested often and proper intuition is essential; these are all skills that were never really developed for me in my secondary education. This sense of unpreparedness can be prevented if students were given a picture of the destination before the beginning of the journey; that is, learning the usage and practicality of the skills before learning the actual skills. Students are never given the real life applications to these subjects, students are left wondering: why are these even subjects worth considering and studying?

Furthermore, I feel like I am just now learning simple Mathematics, as it's just beginning to come together and make sense to me. A great example depicting the lack of coherence are my experiences in Geometry; any 10th grade student may be able to tell you that the area of a circle is $A = \pi r^2$, but I just realized this year in Multivariable Calculus, that the derivative of this area with respect to r is the circumference of said circle, $C = 2\pi r$. Additionally, if we consider the derivative with respect to r of a volume of a sphere $V = \pi r^3$, we can also see that it's the surface area of said sphere $S = 4\pi r^2$. Why is my basic understanding of mathematics just now coming together?

Am I at fault for this, I don't like to think so. Rather, it's the way in which Mathematics is taught in secondary education which is at fault. It's as if students are getting only some of the story, rather than the full story. They're left with gaps in their understanding, and as a result they're unable to see the true purposes behind the subject; as it's not being depicted in its entirety. Students are left in the grey, unsure on what it is that they're even learning. In the head of a young student: there is no coherence, and there is no structure. Mathematics in particular would be a lot less cumbersome and more enjoyable for students if they didn't feel required to lug around a brain full of irrelevant formulas. Is it really any surprise that one of the most dreaded courses amongst students today is Mathematics?

The pedagogy of Mathematics, as well as all other subjects, would benefit from considering my individual experience as I'm sure that there are many others who can relate. If Mathematics were considered more interesting, then more students would find themselves contributing to this realm. This special case of Mathematics is not specific to itself, it is analogous to just about every other subject. If subject material were more coherent and interesting then more students would see subjects for their true values that they have, not just in the classroom setting, but in the world and beyond.

Huda Abushanab: Challenging the Status quo in Undeserved K-12 Schools

To implement a successful system to educate students for conceptual understanding requires a knowledge in pedagogy. Science education programs usually include science content courses, but little attention has been given to the knowledge pertaining to the pedagogies of science content instruction. This is a critical aspect of teaching that is being demanded by students and student advocates in order to improve science education.

From my own experiences, as a child that has moved and attended 14 schools throughout grade school, I saw how different approaches of pedagogies have different effects on students. There were teachers I learned from, but there were others who encouraged me to not only learn, but fully understand the content and concepts, where I still retained the information years after their class. As a student, I don't retain or learn much when a teacher is just up in front of the class lecturing. I learn through a process of trial and error, asking questions, and applying what I learn to solve problems. Problem solving-based learning would let me go through that process to learn, not just the content, but the concept as well. Lectures restrict students' curiosity and limit their ability to explore the subject being taught.

Student-centered and active-learning pedagogies benefit students the most. Problem- based learning is a very effective method, in which basic science concepts are learned, discussed, and applied in a practical, real-world context. A key feature of this pedagogy is that the problems come first and introduce the content. That way, students can learn by exploring the concept and become self-directed learners. Lecture-only methods of teaching that attempt to transfer knowledge directly from instructor to student are ineffective for most students. A lecture only pedagogy is very content-centric, which doesn't engage the students and doesn't allow the students room to explore their thoughts and ideas. Content and concepts of science are more likely retained with a problem based learning approach than a lecture only approach. Teachers need to understand that knowledge is built in the students' mind by the student.

Problem-based teaching caters to every students' needs, especially in a diverse classroom, where each child has a different set of background knowledge. It gives students the space and flexibility to approach problems in a way to best accommodate their understanding. Lecture based teaching should no longer be a main pedagogy in classrooms, especially science and math classes. Flexibility is a key aspect in the pedagogy enforced in classrooms. Students are different and maybe some don't see the relevance in education. A teacher has to be prepared for changes that will accommodate each student. Teaching should not be "one size fits all".

To achieve this, comprehensive programs can be implemented to help teachers from all levels of experience. Teachers may have been teaching for years, successful too, but there is always room for change and improvement. All stakeholders of this education system will benefit. Students will become more driven, passionate, and motivated due to teachers' encouragement and support. With a problem- based pedagogy, grade school students will be better prepared for college because allowing kids to have the freedom to learn their way, makes it easier for them to learn on their own once they are in college with less teacher-student interaction. Teachers will also benefit as well because they'll see that their job was done right. A teacher's main goal is to teach and have the child learn, right? Teachers will also get feedback from parents and administrators, encouraging teachers to do their best. Everyone benefits and succeeds.

Alexiz Gomez: Step it Up

Science itself is a very broad field; there are many disparities between the different fields in science but it is especially evident between science and the public. The disparity not only lies between the flow of information from scientists to the public, but also from the public to scientists. Understanding the way in which science is taught, as well what and how it is taught is crucial to the public because it will either reinforce a disdain or an interest in the scientific field. It is important for scientists to have an idea of how the public thinks and feels about certain issues so that they could address areas of interest. I feel that the public, particularly the youth from elementary to high school, should be taught differently and more in depth because all too often we as scientists discredit the depth to which children can learn. The pedagogy, the method and practice of teaching, of science in schools should be taught in more depth because it is going to be the children who eventually grow to become the broader public. If the children are more educated in science then they will be more likely to gap the bridge between science and the public; this will create an equal flow between scientists informing the public and the public informing scientists.

As a volunteer in a third grade classroom, I have seen the massive potential kids hold; their minds are able to retain and understand much more information than we give them credit for being able to do. I have witnessed that the way teachers are teaching kids, and they are not as effective as they could be. The system in which they teach is saddening because they are on a strict "schedule" to keep all kids on the same track; this type of teaching will eventually suppress the children's curiosity instead of promote it. Why? Because of the schedule they are on doesn't allow for digression into topics that could have real world implications.

I feel that in order to create a teaching environment that students can thrive in, we are going to need to change some of our teaching methods. The teaching methods in America are lackluster in comparison with the teaching methods of other countries. In one study, a man left a computer in a wall of a rural area of India where technology was nowhere to be found. The children, who only speak hindi, became so fascinated and curious with the object that they not only learned how to navigate this English based computer, they also learned how to DNA transfer worked as well as many other things students 5 times their age were learning. The most amazing part is that these children were not just regurgitating material, they were learning and understanding. Through this I see how in America we give the answers out to students so easily that they find no interest in the subjects they are learning. I feel that teaching subjects such as science shouldn't follow a strict schedule, it should encourage curiosity and a style of learning that works best for the student to be most engaged in whatever they're learning.

We can begin this shift in educational instruction by realizing that our education system should not be mechanical and stuffed into a "one size fits all belief system"(Matsui Lecture); we are educating humans, meaning we should be focusing on a system that focuses on the people, valuing each individual's diversity. By replacing our "culture of compliance"(Robinson) with new teaching methods that engage the students. If we can create a form of teaching that sparks curiosity within the students, a form of teaching that recognizes that each student is different and diverse, and that that is a wonderful thing, the students will likely be more involved in the learning system and be excited to gain more understanding in all areas of school (Robinson). By doing this we will also learn to place the same value on the humanities as much as we so the sciences. This is so important because it widens more than the analytical mind, it creates pathways for the innovative mind to roam free which will essentially give students greater perspective and understanding in fields like the sciences.We can begin to make these changes by creating comprehensive programs that help beginner teachers identify and make these adjustments so that we can increase teaching effectiveness. Encouraging "the students' curiosity, individuality, creativity" (Robinson) is how we are going to get them to learn because if you can light the spark of curiosity in a child, they will likely learn because they *want* to learn.

Works Cited:

1. How to Escape Education's Death Valley. Perf. Ken Robinson. Ted Talks, n.d. Web. 07 May 2015.

http://mcbfifteen.tumblr.com/post/118681426390/mcb15-the-pedagogy-of-science

Mike Espino: Teach For American Science

Pedagogy, the method and practice of teaching, is an extremely important topic in science education. Pedagogical techniques should work toward creating a highly educated and well-informed public; in a sense, the pedagogical practices of a society can define the knowledge that society holds. When it comes to science, the value placed on education is high, but the lack of well-trained educators in K-12 schools does not align with the high worth placed on scientific knowledge. In particular, programs like Teach For America (TFA), which are often funded by the federal government, aim to create and train more teachers for underserved school districts. This mission is admirable and aligns well with the value placed on education, but the practices by which TFA teachers are created have clear problems. Those who become part of TFA are given a summer of minimal training in pedagogical theory and technique, and then placed in school districts where the need for teachers is greatest. Indeed, this solves the problem of quantity, but contributes to the problem of quality; because the least experienced educators are placed in K-12 schools where the need for educators with experience is needed the most, the children end up losing. This has grave repercussions in all subjects, but the effects are especially large in science. Through personal experiences it has become clear that in these situations, only students with an innate aptitude for science tend to succeed in such environments. This experience aligns well with Charis Thompson's interpretation of the 'scientific pipeline,' particularly with her critique of directed paths in STEM only for those who succeed early on. Charis argues that the scientific pipeline is built in such a way that it has no re-entry mechanism for students to engage in scientific study if they leave it at any point. As such, I agree with her proposal for a new pipeline with a built in mechanism for re-entry into the sciences. Furthermore, I believe programs such as TFA should have more rigorous training for all educators; it should include specialized sections that will prepare teachers for the rigors of teaching math and science in settings where this is traditionally challenging. This will facilitate the transcendence of TFA from a program that perpetuates the myth of 'low-performing' K-12 STEM schools, to a more well rounded system for creating high quality educators.

http://mcbfifteen.tumblr.com/post/118681426390/mcb15-the-pedagogy-of-science

Ai Hua:

In the paper titled "Female Friendly Science: Including Women in Curricular Content and Pedagogy in Science" published by Penn State University Press, it was projected that "there will be a larger segment of minorities and females" joining the

U.S. workforce; however, the nation will still experience a shortage of scientists and engineers which can tremendously impact the U.S. economy. In response to this prediction, "the National Science Foundation and the Office of Technology Assessment recognized[ed] that science should no longer remain a white male preserve" and "women and men of color must be recruited into the pools of potential scientists in much larger numbers" to alleviate this potential shortage. Since the 1900s, the emphasis of science, its importance, and its impacts began to increase across the United States. The field since then began to open and appeal to women and minority groups for the potential shortage of scientists and engineers in the nation and the quickly increase in immigrant populations.

Now in 2015, we see that more and more individuals from minority groups have been and are going into research as well as pursuing science-related, or STEM, careers since research began to gear away from white male tradition. Despite the effort, the selectiveness of the field has rendered it a rather competitive path, leaving individuals from these minority populations in a less advantageous position and still remaining under-represented in the science field. This has a lot to do with pedagogy of science; in other words, the education or teaching of science, what and how students are taught about science in school. Grade school period is an impactful time period as this is when the majority of students begin to develop their interests in specific subjects or activities; looking from this perspective, the teachers are analogous to scientists and the students the public. Depending on how the students' exposure to science is shaped, this experience may greatly influence their understanding of the role of science in the society and their consideration of the public while performing science later on life. For the discussed reason, our group has focused on analyzing the K-12 education in economically disadvantaged communities from three viewpoints: our current K-12 public education system, our ideal K-12 public education system, and our suggestions on sustaining the ideal system.

While pedagogy should be achieved through the collaboration between teachers, students, parents, and administrators, our current K-12 public education system illustrates a hierarchical model where students are at the bottom of the hierarchy, being heavily influenced by the way the education system is run by the administrators, who are often on the top of this hierarchy. Teachers and parents are often pictured to be in the middle. Although this hierarchy may provide an ordered organization when it comes to decision making, it does not give everyone in the system a fair voice. In addition to this hierarchical organization, the way the majority of teachers are trained does not provide them the necessary tools and knowledge about the complexity of science and how it connects to their daily and the prospective communities that they might serve in the future.

This project has given the opportunity to make such observation as well as become more knowledgeable of how the K-12 public education system is operated. In the process of collecting information for this project, I have learned about the various exams required for someone to become a teacher, such as the California Teaching Credential exam and many other subject-specific exams. Among helpful teaching concepts developed to assist the teaching of science, the concept of Pedagogical Content Knowledge, PCK, is one of the most widely applied. The concept was introduced by Lee Shulman in 1986 expressing the relationship between the knowledge of pedagogy and content that can enhance teaching quality of teachers and learning quality of students, especially those are in the STEM field. "Pedagogical content knowledge is an accumulation of common elements: knowledge of subject matter, knowledge of students and possible misconceptions, knowledge of curricula, [and] knowledge of general pedagogy" (Herr, Ph.D). Many other concepts have also been developed and proven to positively impact the students' mastery of subjects; however, it seems to be almost impossible to find pedagogical concepts that emphasize or even touch on the importance of public understanding of science.

As often discussed in lectures and discussions, my group's findings have repeatedly confirmed our standpoint, that is the current system only provides professional training for teachers yet lacking the comprehensive training, where teachers are trained to encourage students and help them see how science actually connects to their life and that they have a right to voice their opinions in the system as their teachers, parents, and the administrators do. In other words, the way teachers are trained to teach does not focus on collaborating and bilaterally communicating subject materials but rather meeting State's performance standards using techniques to convey their ideas and relay the materials to students. Grade school students are consequently not often exposed or encouraged to think about science from the standpoints of the public, who are affected by research and studies in many different ways. In effort to bridging the gap between science and the public, not only grade school teacher but also students should be informed or aware, if not knowledgeable of this disparity or gap, so that they can carry out more public-oriented research as how it is supposed to be.

Works Cited:

"Pedagogical Content Knowledge." Pedagogical Content Knowledge. N.p., n.d. Web.

Rosser, Sue V. "Female Friendly Science: Including Women in Curricular Content and Pedagogy in Science." JSTOR. Penn State University Press, n.d. Web.

Koehler, Matthew. "Pedagogical Content Knowledge." Pedagogical Content Knowledge. TPACK, 13 May 2011. Web.

"Pedagogical Content Knowledge in Science Teaching." The Sourcebook for Teaching Science. California State University, Northridege, n.d. Web.

Part 2: Future of Science and Sustainability of Change

We plan to carefully examine how the continuous development of science affects our ethics in the future.

Michelle Guillen: Future of Science and Sustainability of Change

Science is advancing rapidly. As science continues to advance, there are several ethical concerns about the accessibility of research. Accessibility to primary resources has been hard to obtain for people outside of education. In terms of accessibility, it is common that only faculty and students in higher education are able to obtain the hard evidence present in journals. This creates a gap between those that are able to obtain knowledge about scientific advancements and those that cannot. While this disconnect has been somewhat balanced by the increase in science being displaced in popular media, there is still this ability for opinions about science to be formed by incorrect or opinion based popular media articles. The future of science consists of rapid advancement and the necessity of transparency between scientists and the world.

The increase in technology has created a modern scientific revolution. However, there is this lack of communication from scientists about this advancement. Recently, Dr. Canavero has announced his plans to do the first head transplant surgery. In China, scientists have reported editing the genomes of human embryos. Even the manner in which clinical trials are run illustrate the lack of ethical concern between the science community and the general public. The popular science articles that discuss the advancements in science rarely discuss the manner in which these advancements are run. They merely reflect decisions already made or done by scientists in proceeding with their research. There is usually little room for the people to argue against the science going on. In order for change in science to be sustainable, there needs to be a bigger engagement between people and scientists. While this is mediated through popular science media, it does not bring the engagement that is necessary to make accurate decisions about science. Regardless of whether the public is involved, science will continue to advance. However, this change is only sustainable with the ability to have the public access evidence to make science ethical. In terms of increasing transparency, it has already been noted that those involved with higher education institutions have the means of obtaining primary resources. Therefore, the first step in having open communication about science advancement is educating students, which are the future leaders of the world, about how to access these articles. This would be a more direct way of having a portion of the nation's voters informed on research aside from going through the process of making all journals open access. Overall, science will advance but this will only be funded if the public is being informed about the advancements rather than finding out about them after the fact.

Andreas Rodriguez: How are controversial scientific decisions decided on?

We are at a point in time where technology is advancing at an incredible pace; some may even say it is advancing too fast. New techniques are popping up all the time but who is to say whether or not they are ethical or safe. I am first going into the past to see how a well-known controversial abortion case was handled. In 1973 the Supreme Court ruled in the case *Roe vs. Wade* that women should have the choice to get an abortion as long as it was restricted to the first trimester of pregnancy. This decision has been tirelessly debated over the years and still continues to be a hot topic to this day. In this case and in many similar to it, questions of ethical conduct are determined once they reach a court, and are decided on once someone feels that they have been wronged. In this sense the public is represented in the form of a jury, and if the decision that is decided on in does not sit well with the general public then I am sure that eventually it will return to court once more. However, science experiments are happening fast, and who is responsible for monitoring scientists? In a recent article called "Critics Lash Out At Chinese Scientists Who Edited DNA In Human Embryos" by Rob Stein we see this exact problem. Experiments have been done on human embryos, an action that has been frowned upon by the scientific community as well as by the majority of the world. The article illuminate some of the regulation processes are out there, for example, "watchdog groups", such as Center for Genetics and Society, monitor some of the new age technologies and experiments that are popping up. Other forms of ethical regulations come from fellow scientists, as well as world organizations, who disapprove of human experiments. However, what if these experiments bring us positive results, could it convince other scientists that it is time to begin experiments with human embryos? Will the rules of the game change? Questions such as these bring up very controversial views as well as a need for more public involvement in current scientific decisions. Will we play with our human genome, the very core of our humanity, for the sake of science?

Info graphic: http://mcbfifteen.tumblr.com/post/118668978980

Video/Audio: http://mcbfifteen.tumblr.com/post/118670624685/mcb-15-final-project-designer-babies

Works Cited:

Stein, Rob. "Critics Lash Out At Chinese Scientists Who Edited DNA In Human Embryos."

NPR. NPR, 23 Apr. 2015. Web. 12 May 2015.

Grace Ho Jung Kim: Designer Babies: Ethicality

The beauty of being human is that we live with imperfections and we come to embrace our flaws. If one were born perfect, whatever the standard for "perfection" may be, life for them would be tasteless and dull. Although I do believe that technological and biological advancement in society is essential for progress, giving men the power to alter life is alarming. Thus, despite the positive effects that research for designer babies may bring to our future, I believe that designing our own perfect life is morally and ethically unjust.

The process of creating a designer baby, along with its concept, is morally disturbing. Before the beginning of the 21st century, genetic manipulation has been shown through science-fiction publications. However, with significantly rapid biological and technological advancements, such imagination has become a potential reality. Designer baby is a genetically engineered "in vitro for specially selected traits, which can vary from lower disease-risk to gender selection" (Ly). Such an idea seems very polished on the outside; however, the procedure of creating one through In Vitro Fertilization (IVF) is questionable. There are three steps to this process. The first step is to choose the type of sperm that will fertilize an egg. Second step is to secure it in a test tube outside of the mother's body and use advanced reproductive technologies in a laboratory to modify embryos, providing the option to choose desirable or cosmetic characteristics for the baby. Last but not least, the third step is to implant the modified embryo back into the mother's body. I believe that the term "modify" is something that should not be used when discussing a human life. Simply imagining that the laboratory would, in the future, become a place where every life would be altered and sent back to society, is unpleasant and frightening.

Furthermore, humans are not experiments. Conducting a research and experimental procedures on mice cannot guarantee that it will succeed on humans as well. Compared to other researches completed with mice, creating a designer baby is on a completely separate scenario. This experiment deals with changing the genes and generating a "perfect" human being. Sooner or later, the experiment will have to extend to using human embryos in order to test its potential on humans. However, experimenting on people is appalling and gruesome. Thus, no matter which route the process may take, it will be unethical and undesirable.

Furthermore, a life is something that we cannot create nor can change to fit our desires. Who would set the standard for what is "desirable" or "cosmetic?" Certainly, the embryo does not have any say in a process that is doomed to transform his or her life permanently. I fully comprehend that this advancement allows doctors and parents to screen embryos for any genetic disorders and select the healthy ones. Such techniques allow "doctors and parents reduce the chance that a child will be born with a genetic disorder" ("What Is a Designer Baby?"). However, we need to carefully examine the costs of such innovation. Throughout the years, the definition of beauty has changed numerous times. In South Korea, parents give their children plastic surgery, as a high school graduation gift because acquiring European features, such as double eyelids, higher nose, and sharper jaw line, has become a standard of beauty (Chung). Despite such obsession to our set standards of beauty, the distinction between plastic surgery and designer babies is that the former is "adding" to a natural, while the latter is changing the natural all together. Plastic surgery "reforms" the parts of which the person desires. In contrast, designer babies themselves have no choice or say in becoming one and will possess superficiality. The joy of seeing one's child beautifully grow with features similar to their parents is a blessing and a gift of god. Additionally, being able to see such genes shine through a growth of a child has been a part of being human since creation of the earth. Yet, resting the power of life in the hands of men seem disastrous and disrupting the natural order of the world.

To an extent, I would be in support of genetic modifications if that meant that the baby would grow up free of genetic disease that he or she may have been initially predisposition to. However, I am not in support of creating biologically advanced babies. In addition to many ethical concerns, there are economical worries involved. The process of creating a designer baby would, initially, only be available to the very rich who can afford this treatment, thus leading to a class divide that is not only rooted in money and status, but in our very own DNA. I believe this would be a step towards what Adolf Hitler wanted to build, a "master race." Soon, designer babies would become the norm of the society, where their perfection would also seem mundane. I believe that human greed is incredibly powerful, thus even after we have made great steps towards perfecting humans, there will be no end to it.

Personally, as a Christian, I believe that God is the being who has the authority to create, foster, develop, and destroy all living things, including life. In establishing designer babies, it is as so we are playing God. Furthermore, as a Political Science and Ethnic Studies major, I cannot help but sympathize with all the non-ethical parts of this research. As I stated in the beginning, part of being human is that we live with imperfections and we come to love our flaws naturally. There is a reason for everything. However with designer babies, there would no longer be a purpose of natural selection and genetics. Again a part being who we are is that we foster ourselves into respectable and mature human beings by learning how to deal with our flaws and accepting the shortcomings of others. With perfectly designed human beings, the concept of, acceptance and maturity will take on a different form, if not disappear all together.

Apart from all negativity and ethical arguments, however, it is important to note that such advancement when used in acceptable situations and regulated heavily, will bring great progress to humanity. Like I mentioned before, there are benefits to genetic alterations. We will be able to scan our embryos and eliminate any potential diseases. However, when this extreme power is placed in the hands of the wrong group, it can easily be turned into a manipulating power. Therefore, regulations for the research and creation of designer babies must be regulated by a group of trustable scientists and United Nations. Genetic research occurs all over the world and we really need to firmly establish and enforce universal rules over human genetic experimentation. Strict guidelines will help in managing the project. This way, scientists will be able to relay information to United Nations for the worldwide organization to spread the knowledge in a simpler, yet comprehensive manner. Such efforts will be a step towards lessening the gap between the public and the scientists.

Designer babies can be used in numerous ways, such as for beauty, health, or governmental usages. However, it is undeniable that the process of creating a life in a laboratory is unethical and contains the potential threat of altering humanity forever. When regulated heavily not only by a group of scientists, but also by a larger, governmental organization, the project will bring great progress in helping to make our world a better place. The future of science is bright in my eyes.

Please take a look at our info graphic and audio recording here:

http://mcbfifteen.tumblr.com/post/118668978980

http://mcbfifteen.tumblr.com/post/118670624685/mcb-15-final-project-designer-babies

Work Cited

Ly, Sarah. "The Embryo Project Encyclopedia." Ethics of Designer Babies. N.p., n.d. Web. 09 May 2015.

Marx, Patricia. "The World Capital of Plastic Surgery." The New Yorker. N.p., n.d. Web. 09 May 2015.

"What Is a Designer Baby?" What Is a Designer Baby? Bionet Online, n.d. Web. 09 May 2015.

Seoiyoung Ahn: The Development of Science and Bioethics

Astronomy Professor Fillippenko at UC Berkeley said that "if the 20th century was the century of physics, then 21st century is and will be the century of biology." Nowadays, the expansion of field of science, especially biology, is accelerating at an unprecedented rate; everyday, we face new discovery in biology that will bring about a possible effect on our lives. For example, professor Doudna at UC Berkeley surprised the world as she successfully conducted a research on an enzyme, called CRISPR/Cas9 Nuclease. Her study on this was ground-breaking because this enzyme was capable of inserting a certain mutation in a DNA region where we want to insert. Then, this research recently raised the possibility that perfect" baby can be designed before they are born by modifying and manipulating the baby's genetic information—designer baby

Although this seems very appealing and amazing, there are issues that are closely concerned with bioethics of "designer babies." If scientists become capable of designing a baby, could this method be justified bioethically? How about the effect that it would bring about to our society? Don't you think that the rich, who can afford this, will become smarter and thus richer in the future, whereas the life conditions of the poor will get even more miserable, thus widening the gap between the rich and the poor. To learn an accurate picture of how people around me thought about this designer baby, I obtained thoughts regarding this from 40 people. Then, it was not surprising that about a half of my friends whom I asked said that designer baby is not bioethically right. However, another half said that said that designer baby is not bioehtically wrong at all. And, this survey hinted the diverging thoughts of many people toward the designer baby. One of interesting that was made by my friend was that "designer babies" are not bioethically wrong at all, because it does not need and require killing any type of living creature. Usually, biological research sacrifices animals to bring about positive benefit for human; killing a living creature to bring about benefit to human beings is bioethically wrong, he claims. However, designer baby does not require any sacrifice. Nobody—specifically, no babies—will be killed, but scientists and/or doctors are simply "modifying" genetic material to simply introduce changes to these babies.

Bioethics and biological research are something that can never be separated; wherever there is biological research that is going on, there will also always be issues of bioethics. One of the best way to resolve the conflict between biological research and bioethics would be set up specific criteria that will determine whether a certain biological research is bioethically wrong and right; currently, the world—including scientist community, population of general people, the congress—are devoid of criteria to determine the bioethics of biological research. With this urgent need to do so comes another question: Who should determine such criteria for bioethics? Should be scientists' community, general population of people, or the congress? People's thoughts about this seem to be divergent as well and therefore places a dilemma towards the goal of resolving bioethics and biological research. If this conflict persists on, the issue of bioethics may be ignored in the future, as there is no way that biological research will stop to conserve bioethics. If such things, the world and the community of biology will be in chaos, reaching appoint where scientists may be doing research or experiments on human beings by killing human beings.

Thus, we do need to find a compromise between bioethics and the biological research; biologists will continue research and those research may, one day, become unquestionably bioethically wrong—bioethics is a only means to prevent from that happening. In order to prevent this from happening, we—including scientist community, the congress, the government, and the people in the world—should take a step to resolve this conflict. Once the issue of bioethics is resolved, the biological field will be much more beautiful and will be able to bring about "right" and "appropriate" improvements for human beings.

Shahil Zhangada: Equipment Used for "Designer Babies"

Technology is on a rampant rise. New information has been processed where software can now indicate whether or not a child has Down Syndrome while still in the mother's womb—in fact, technology is at the point where a mother could alter her baby's physical make-up while still in her womb. The latter application of technology is commonly attributed as "Designer babies""—a process where consumers are able to select traits for their babies, essentially creating people with characteristics deemed viable by their parents. Though this issue is considered controversial, the equipment used in creating a "designer" baby is not; the machinery used has an objective function and it is the purpose of such equipment that provides the ethical issues.

The process for genetically modifying one's baby has evolved over the years—and now, provides a consumer-friendly way of picking which genes to keep and which ones to modify. In fact, according to *BioNet*, babies that are kept through InVitro Fertilization (IVF) outside of the mother's womb are able to have their genes modified, as they are able to be screened for genetic disorders (http://www.bionetonline.org/english/content/db_cont1.htm). At the moment, there are three prominent methods for editing embryos—the first is through stem cells from a sibling, as stem cells can function as any other cell when modified, the second is through analysis of sperm (rather than embryos) to screen for sex-chromosomal issues, and the last is through the usual PGD (Pre-Implantation Genetic Diagnosis), where doctors are able to screen genes and select the optimal genes to place back into the mother's womb. Similarly, when it comes to gene editing, a common process involves the CRISPR/Cas9. In nature, microbes use it to immunize their DNA against viruses, by snipping out distinctive sections of viral DNA and pasting them into their own genome. In 2012, biologists figured out they could use the system to edit the genes of any organism (http://www.wired.com/2015/05/read-freak-gene-edited-superbabies/). The different methods of choosing genes to specify a parent's wish and desires may prove controversial, but the technology used to achieve such a feat prove to be quite objective and sophisticated. The equipment displays a progression in technology, which agrees with the notion of a rapidly growing technological sector. The issue with the technology is the bioethics behind the applications of the technology. Will the equipment or machinery be used to perform ethical duties? Or will the technology be used for immoral use?

When it comes to "designer babies," researchers have indicated a growing concern in popular opinion. Essentially, media outlets such as the news or online news sources have been able to paint this specific application of "genetic editing" or genome editing in negative and positive light—often times, it is often up to the individual to make this distinction on their own. However, there is little debate over the processes used to edit genes—genome editing has been around for a while and has positive benefits. From genetically modified organisms providing cheaper and more accessible food to genetic screening of human embryos to detect genetic disorders or even determining the sex of a child, the processes used to create a "designer baby" have positive connotations and applications. But why is it that when the issue of "designer babies" arise, people begin to bash technology and the processes used to modify genes?

The issue here is the "gap" in knowledge of the issue. Due to unequal access to scholarly articles or news on modern technology, the different strata of people have varying knowledge on certain topics and so, have different tendencies when reacting to opinions on these topics. With varying knowledge on issues, the result is an uneven surface of opinions and different comprehensions of certain definitions. The distinction between process and purpose must be distinguished when it comes to discussions of ethics. The technology behind any application is objective but the purpose of that application is adherent to ethical debate—and that is the same with designer babies. The technology behind it is not unethical but the way that technology is used is open for debate and discussion.

Info graphic: http://mcbfifteen.tumblr.com/post/118668978980

Video/Audio: http://mcbfifteen.tumblr.com/post/118670624685/mcb-15-final-project-designer-babies

Bryan Huang: Ethical implications of technology in science

Technology has transformed the way that we practice science greatly. However, as we push the borders of what is possible, we've also opened up a plethora of ethical issues that we must carefully think about. Technology is definitely changing the world in amazing ways, from how it has connected our world together, to all the various inventions and ideas that have made our world what it is today. This is no different for scientists and researchers either - technology has been able to connect researchers in all sorts of different fields together, and technology has enabled discovering and experimenting things at an incredible pace. However, with great power comes great responsibility. As technology pushes science forward, new ethical questions are constantly brought to light. I feel as if those who practice science, though they might be excited and enthusiastic about their work, do not put as much thought into the ethics involved in it as they should. Ethics is tied very closely to those outside this research community - I think part of why we have this idea of "ethics" is because those both inside and outside active research question the values and morals of what is being researched and done. In a way it is reflective of the feedback for those both within and outside of the scientific community. It's important to consider the ethical implications of what we do, because this is one of our responsibilities as being part of our society.

For example, recently researchers were able to create the first genetically modified embryo in the world. While the intention behind it was good - to cut out a certain part of DNA that causes a life threatening disease, concerns have been raised about whether it is ethical to modify embryos. This is very closely related to issues such as cloning and stem cells. Whichever side you decide to argue for, I think it's important to keep an open dialogue over new practices. What we consider to be right or wrong is constantly changing - as an example related to technology, copyright laws have changed a lot. What was considered breaking copyright in the past is now is seen as encouragement to promote sharing. Similarly, the notion of ethics is also constantly changing. What's important is that we should always think about the implications of what we do. This extends beyond science and technology. As long as we're able to maintain an open dialogue, and do things that are in the best interest of our world as a whole, then a lot of the barriers to understanding each other can be broken.

Info graphic: http://mcbfifteen.tumblr.com/post/118668978980

Video/Audio:http://mcbfifteen.tumblr.com/post/118670624685/mcb-15-final-project-designer-babies

Works Cited:

http://theconversation.com/worlds-first-genetically-modified-human-embryo-raises-ethical-concerns-40766

http://listverse.com/2013/07/12/10-recent-scientific-advances-that-signal-the-future/

http://undsci.berkeley.edu/article/0_0_0/whathassciencedone_03

Final Thoughts

When a student surpasses the barriers of racism, sexism, ableism, classism, among countless others, and gains admission into a university, why is it that this student continues to be subjected to these barriers while studying for classes, working multiple jobs, and volunteering for the community?

A classroom that strives to understand its students, rather than having students try and understand the sliver of knowledge that they are tested on to soon forget, is just another format we can consider as we change what it means to work and study at an institution of higher education.

Is higher education still mastery of classic research and texts, or is it development of community? Which affects our behavior more: content or experience?

Investigating how science informs the facets of life is just a part of this book. How can we unsimplify "the public" and broaden perspectives on "science"? Just ask the millions of students majoring in STEM, or choosing not to, what their stories are. Such experiences will undoubtedly shift the values of our practice.

Thank you.

www.ingramcontent.com/pod-product-compliance
Lightning Source LLC
Chambersburg PA
CBHW080905170526
45158CB00008B/1994

* 9 7 8 1 3 6 5 0 6 5 3 6 1 *